POLYPHONIC COMPOSITION

An introduction to the art of composing vocal counterpoint in the sixteenth-century style

by

OWEN SWINDALE

Music Department
OXFORD UNIVERSITY PRESS
44 Conduit Street, London W1R 0DE

First published 1962
Re-printed 1964, 1966, 1969,
1972, 1975, and 1977

Reproduced and printed by
Halstan & Co. Ltd., Amersham, Bucks., England

CONTENTS

		Page
Introduction		i
Note to the student		iv
CHAPTER 1	*Plainsong*	1
CHAPTER 2	*First species*	4
CHAPTER 3	*Second species: the passing note*	11
CHAPTER 4	*First and second species mixed*	16
CHAPTER 5	*Third species: melodic decoration*	24
CHAPTER 6	*Fourth species: the dissonant syncopation or suspension*	82
CHAPTER 7	*The synthesis: fifth species*	35
CHAPTER 8	*Fifth species with a cantus firmus*	44
CHAPTER 9	*The modal system in the 16th century*	50
CHAPTER 10	*The two-part fugal style: counterpoint without a cantus firmus*	55
CHAPTER 11	*The homophonic style in three parts*	73
CHAPTER 12	*Fantasia on a plainsong cantus firmus*	82
CHAPTER 13	*Three-part fugal style*	86
CHAPTER 14	*The three-part motet*	96
CHAPTER 15	*Four-part counterpoint*	114
Appendix One	*Synopsis of rules*	123
Appendix Two	*Semibreve time and crotchet time*	125
Appendix Three	*Double counterpoint*	129
Appendix Four	*Plainsong canti fermi*	135
Index		137

INTRODUCTION

There are no fixed and eternal laws of counterpoint; for counterpoint is in the last analysis simply composition, and composition is what composers do. And yet there are rules for the game even if we invent them ourselves. Indeed we must, for even the most elementary piece of musical construction implies selection and rejection of many alternatives.

At its best the training of a developing musical consciousness means the uncovering of the manifold possibilities of manipulating intervals, with as little bias as is reasonably possible towards one particular set of possibilities. For real progress one must begin with the thorough exploration of diatonic intervals - and for preference without being tied to a major-minor key system. For anyone willing to try an exhaustive inquiry into what is still possible in a purely diatonic set of scales there is a shock waiting - not a hundredth part of the potential combinations of intervals, vertical or horizontal, has been used, or even tried. The advantage to a composer of tonal values that are *aurally identifiable* is absurdly overlooked at the present time.

The advantages of studying the polyphonic school of composition are these:

(1) We can begin at the beginning: with our attention focussed closely on the melodic line, and on the sounds of intervals in two parts.

(2) We can work on a smaller scale: sixteenth-century music does not require extension as one of its elements. Bach in eight bars is an impossible task, though this is a constant demand in academic work.

(3) In the constantly changing style of different periods the sixteenth-century is a temporary plateau of settled taste and agreed standards. For once style and content are in accord, the vertical and horizontal aspects reconciled. If we are unaware of this, the highest point of equilibrium reached by music, we are in the position of a philosopher who neglects the study of St. Thomas Aquinas.

(4) We can be free of the weary round of I-IV-II-V-I chords which in harmony only develop to chromatic versions of the same progression. The study of conventional four-part harmony has some serious defects, although useful enough for gaining an insight into eighteenth and nineteenth-century compositions. At the present time, when music is again being remoulded in a more linear pattern, a more harmonically uncommitted style would seem especially appropriate for study.

(5) The step from sixteenth-century counterpoint to the harmonic counterpoint of Bach and Handel is a natural and reasonable one - nothing need be unlearned, and the flexible, extended melodic line that is the prime basis of movement in the Baroque period has already been studied. Teachers who attempt the 'Bach style' without a previous study of 'strict counterpoint' soon realize that the greatest difficulty is encountered in teaching rhythm and the balance between accent, syncopation, and rise and fall of pitch.

(6) The beauties of polyphonic music are not on the surface, but are appreciated by the conjunction of intellect, emotion, and physical action: or in other words, by study of the score, by hearing, and by singing. And if this is not a perfect training in musical aesthetics, I have yet to hear of a better.

The retention of the 'Five Species' as a method of teaching deserves, perhaps, a word of explanation. Research into the music of previous centuries has shown that the method of teaching 'strict counterpoint' in general use was based on a misunderstanding of the style of Palestrina. Originating with Fux, whose (in many ways excellent) *Gradus ad Parnassum* seems to have been the original cause of the trouble, the error crept from one theorist to another. In clearing the undergrowth, R.O. Morris and his followers did much fine work; but they failed to replant the tree. It seems clearer now than it did to Morris, that we can have the best of both worlds - the species *and* the true sixteenth-century style.

So I retain the species; they are hallowed with age (composition was certainly taught in the sixteenth-century by this means) and they can be trusted to produce some results from the dullest of students. On the other hand, the genuinely brilliant will benefit by exploring all the possible moves available with the most restricted means. And it is perhaps the only method which enables a student to make an early start in counterpoint, in itself a sufficient justification.

It is for the teacher to decide whether or not to work throughout the first section with modal cantus firmi. On the principal of not overloading the beginner with too many new facts I have deferred the introduction of the modal scales until the last possible moment. However, they could well be introduced at any earlier point, and certainly there is, or should be, nothing difficult or intimidating about them to the average student.

With this point in mind it is important that the historical background should be filled in rather sooner than later. The hearing of some recordings of plainsong is indispensable while working Chapter One, and the study of the Renaissance period should be well under way if not complete by the time Chapter Four is being worked.

In any case some group singing must be a sine qua non, and while an organized choir is probably the least desirable way of becoming acquainted with this music, it may well be the only practical approach to the problem. The ideal is to make copies of the music available, and let the student do the rest.

It is perhaps worthwhile pointing out that the examples in this book should and could be in constant use as exercises in score-playing at the keyboard. Also, that there is no reason why sixteenth-century music should always be sung if string and wind instruments are available: polyphonic music sounds at its most attractive when played, and it makes an admirable study in the technique of chamber music playing.

The use of C clefs has been kept to a minimum, since I can see no good purpose in making the examples difficult to read. The complete mastery of the alto clef is essential to any musician, and since it admirably bridges the gap between treble and bass it has been used consistently for that purpose. The increasing use of octave transposition for tenor parts I consider deplorable except where it serves a simple utilitarian purpose in a vocal score. When it appears in editions meant to be read in score and intended for musicians and scholars, it is a hindrance.

A similar modern improvement that serves no purpose whatsoever is the halving of time values. This can be confusing to the student and since the different treatment of the various notes values is so important it is better to avoid the use of such editions completely until the style is really familiar.

My thanks are due to the many friends and colleagues who at various times have provided criticism and encouragement, and in particular to Professor Sidney Newman and to Dr. Hans Gal who laid the technical foundations, and to Walter Cairns who guided my foot-steps in preparing for publication.

And finally I must acknowledge my debt to the great musicologists of this age of musicology - to Knud Jeppesen, Willi Apel, Alfred Einstein, and the many others who have made possible a new approach to this subject.

Glasgow, Spring, 1961. Owen Swindale.

Note to the Student

What is taught is first observed: so what is learned must also be first observed. The student will undoubtedly be heartily tired of the injunction to 'study the example below' by the time the first few chapters have been worked: nevertheless it is the kernel of the whole matter. These examples must be read through, played through, and sung through, until not one vestige of strangeness is left to them. Only the mechanics of the style can be taught by means of words: the thing itself, the subtle understanding that a passage looks and sounds 'right', this can only be 'learned by ear'.

However, rules and recommendations must be formulated, if only to save the student the trouble of finding them out for himself. For the sake of clarity they have been stated in as general and dogmatic a fashion as possible, and it must not be a cause for bewilderment if isolated examples can be found which contradict the text. In general I have made it a rule not to exceed the limits of technique set up by Palestrina - while, however, not restricting my illustrations to that composer.

What I have aimed at is what might be called the 'normal style'. The deviations from it will be best understood against a firm knowledge of what is orthodox, and when, as is to be hoped, the student goes on to the study of keyboard music, secular music, and the music of other countries (of which England is not the least important) he will find that although local dialects and idioms vary greatly the basic technical foundation which he has learnt is a 'lingua franca' which is spoken in many lands and is valid as well in the tavern as in the church.

CHAPTER ONE

PLAINSONG

Plainsong, or Gregorian Chant, which is the firm foundation of all our modern art of music, reached maturity in the sixth century A.D. It is thus by far the oldest music that still gives us pleasure. Anyone who has heard unaccompanied plainsong well sung will not need to be reminded of the incomparable dignity and solemnity of its effect, withdrawing the mind from the rush and vulgarity of everyday life into a world of calmness and contemplation.

To appreciate fully this most perfect of arts we must forget the exciting and emotional colours of modern music and learn to take pleasure in a subtle and refined expression of feeling that is not less great because it is on a small scale. Here are four plainsong melodies. Notice the natural and lovely curve of the phrases; the free rhythm, unrestricted by bar-line or time signature, and that the scales, or "modes" though always strictly diatonic (with the exception of the occasional B♭) have an interesting variety of possibilities; more perhaps than our modern major and minor keys.

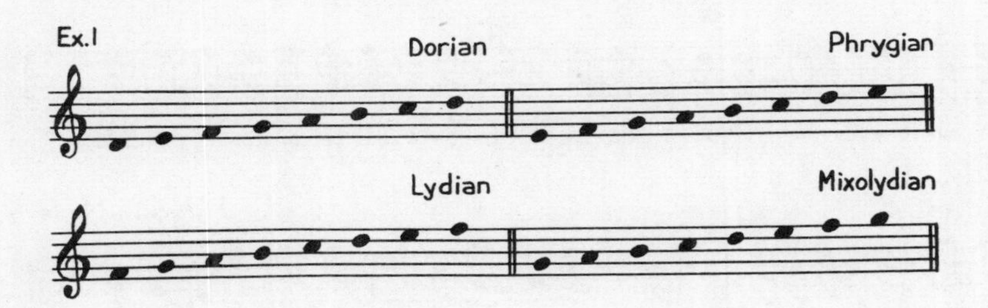

Ex.1 Listen to any available recordings of plainsong.

Ex.2 Sing through these melodies: memorize at least two of them.

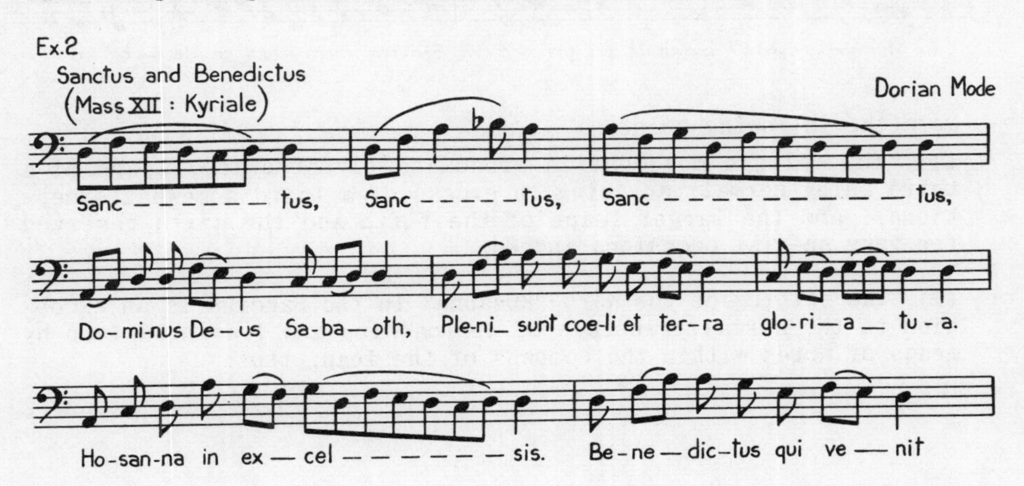

Note the following points:

(1) The melodic movement is predominantly stepwise, leaps of a third being normal, anything in excess of a third somewhat exceptional, and the larger leaps of the fifth and the sixth reserved for very special occasions indeed.

(2) The setting of the word "Hosanna" in the Sanctus is an exception to the normal principle of approaching and leaving a leap by means of notes within the compass of the leap, thus:

(3) The compass of the melody never exceeds an octave; it may lie between the tonic and its octave ("authentic" form) or between the dominant and its octave ("plagal" form).

(4) The note B may be flattened in order to avoid the tritone*, as in the second phrase of the Sanctus. It must be said, though, that the frequent use of B♭ will destroy the individual qualities of the mode, and that the original makers of plainsong seem to have enjoyed the tritone sound when treated with proper artistry. See the delicate and undoubtedly intentional effect of the B against the F in "Lux de Luce", and even more openly in "Claris conjubila".

(5) The rhythm moves in equal note values but the phrases close on notes of double value. The word-setting may be almost syllabic (as in "Verbum caro") or it may contain more or less elaborate "melismata", that is, several notes sung to the same syllable; and in either case there is no regular "beat" - the words govern the accentuation of the music and hence there is a constant shift between duple and triple rhythm.

Ex. 3 Make a plainsong setting of these words:

Agnus Dei, qui tollis peccata mundi, miserere nobis

in the Dorian, Phrygian, and Mixolydian modes.

---

*TRITONE: three consecutive steps of a tone, making up an augmented fourth - F to B. The most awkward interval to sing and therefore one that is carefully avoided either directly

or in outline

in most unaccompanied vocal music. The same goes for its inversion, the diminished fifth, to a rather lesser extent.

CHAPTER TWO

FIRST SPECIES

Monody (a single melodic line) retained its supremacy in music for close on a thousand years, but from the tenth century onward experiments were constantly being made in combining two or more melodies together. As the art and technique of music developed, gradually acquiring rhythmic variety and harmonic coherence, so the names of great composers began to appear - Machaut, Dunstable, Dufay, Ockeghem, Josquin des Pres, and many others. Each expressed in his own unique way something completely personal with the material to hand, and each at the same time refined a little on the work of his predecessors, so that one may say that "without such a composer music would have been different" - not less good, perhaps, but certainly different.

Finally, in the sixteenth century, music reached a very fair degree of perfection, within its own limits (we need perhaps to be reminded that all art must be bound within the barriers it sets itself), and from all over Europe there was a veritable torrent of technically perfect and often inspired works. The language was common, although there were local dialects, and the indisputably great masters Lassus, Victoria, and Palestrina, amongst many others, are so consistent in their use of it that we can formulate rules based on their style with very considerable exactness.

Study the following Kyrie in detail. Sing or play it.

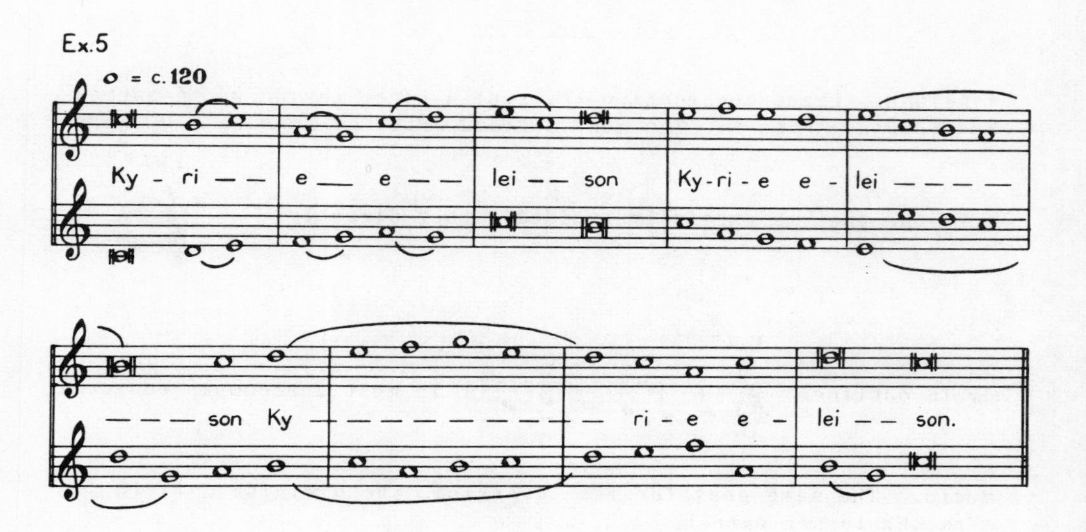

In first species composition the following rules are kept:

Melody

(1) As in the plainsong examples this is mainly stepwise. The leap of an octave may be made, but neither the major nor the minor seventh, nor the major sixth. The minor sixth is best used as an upward leap. All augmented and diminished intervals are strictly forbidden as leaps, and the tritone outline (c.f. Ex. 4(b)) should be avoided.

(2) Any leap is best followed by a step in the opposite direction; but it may be followed by a further leap in the opposite direction, or in the same direction provided that the two leaps do not add up to a major seventh or to more than an octave.

Harmony

(1) Only consonant intervals may be used. These are:

Unison
Third, major and minor
Perfect fifth
Sixth, major and minor
Octave
Tenth
Twelfth, etc.

The spacing should only rarely exceed a twelfth.

(2) Apart from the obvious dissonances (seconds and sevenths), it is wise to be on guard against the intervals of the **fourth** and **diminished fifth**, both of which look innocent on paper but are in fact discordant*.

(3) Parallel fifths and octaves are forbidden since they destroy the independence of the part movement.

(4) Notice the thin and unsatisfactory sound of these part movements:

*The fourth, from an acoustical point of view, is certainly concordant in a fairly high register; but in a low register or if extended to the interval of the 11th or 18th is manifestly discordant. In the case of the diminished fifth the ear of a present day musician interprets it as part of a dominant seventh, a chord which was not part of the sixteenth-century language.

Ex.6

These are "exposed fifths" and "exposed octaves" * and are caused by moving both parts *in the same direction* to an octave or fifth.

The following recommendations should be considered.

(1) Be aware of every interval you write; if it helps you, number each interval:

Ex.7

1 3 6 6 10 8 5 10

(2) Think in contrary motion.

(3) Be sure that you hear what you are writing. At the very least you must be able to hear one line at a time and if you cannot hear both, a little more ear-training practice is indicated. Do not use the piano while working exercises, but never fail to test your work at the piano.

(4) Be clear about the implied harmony. Smoothness of movement and independence of parts are more important than writing 'strong' progressions, but be sure that you know what the chords could be if you were adding a third part.

Rhythm

As you will have seen in the examples of plainsong, the restriction of note values does not mean rhythmic monotony - far from it! Although we now use bar-lines to divide the notes into groups of four the bar-line need not indicate a strong beat. If the word-rhythm demands it, the accent can fall at any point - see Ex.5, bar 6, last beat.

* Also known as "covered" or "hidden" fifths and octaves.

Breves will be most naturally used at points of cadence at the end of a phrase. If they are used elsewhere, care must be taken that they do not disturb the rhythmic flow.

The tempo will be lively - in spite of its "white" appearance on the page a great deal of sixteenth-century music is fast moving.

Ex. 8 shows some of the possibilities in the minor key.* Modulation to the relative major is permitted, since it involves no new accidentals; in general, a completely diatonic style is recommended, since modulation as we know it is not a part of sixteenth-century practice.

Ex. 1. Add a counterpoint below the given cantus firmus. (The part of an exercise which is given is known as a "cantus firmus", plural 'cantus firmi', and was originally a plainsong used as the basis of a composition. The added part is known as the "counterpoint").

---

*Although it would be more correct to work in the modal scales it is perhaps wiser to restrict the number of facts to be learnt at this stage. The more advanced student may refer first to Chapter 9 and then work throughout with modal cantus firmi.

Ex.2 Write out the cantus firmi an octave lower and add a counterpoint above.

Imitation in First Species

As may be seen in the next example, "imitation" is the entry of parts at different moments in time, using the same thematic material. The second, or "following voice", may enter at a different pitch.

The advantages of imitation are: (a) the two parts are heard more clearly, each with its own independent life; and (b) the two parts are bound into a close thematic connection.

The use of imitation involves the use of rests; conversely it follows that if a part rests it should re-enter with thematic material - i.e. a phrase that has been or will be heard in the other voice. Both parts may not rest at the same moment.

At this stage it is unnecessary that imitations should be strict (i.e. with every interval exactly reproduced), and there is no need to imitate more than three or four notes.

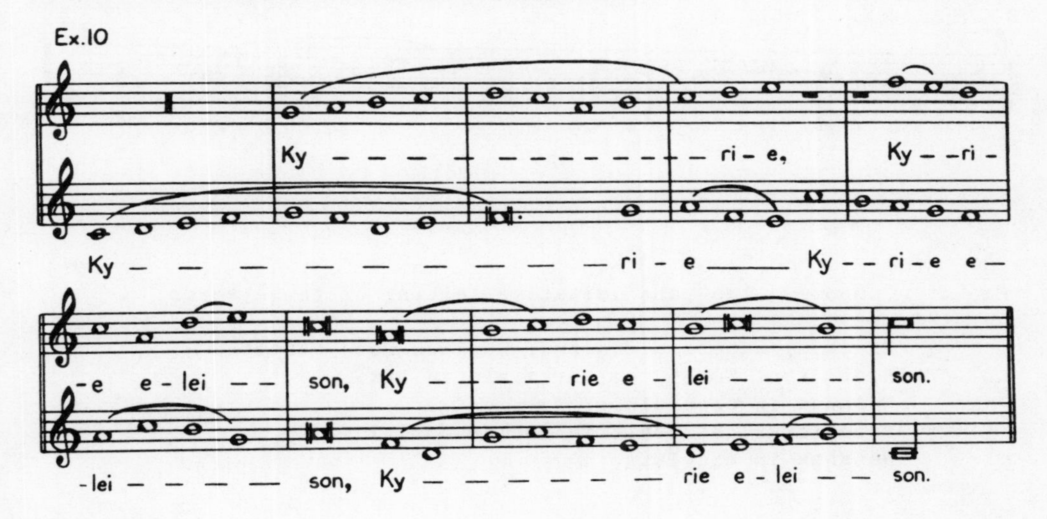

Ex. 3. Complete the following Kyrie, using imitation where appropriate.

Ex. 4. Compose a two-part imitative setting of these words:

Sanctus, Sanctus, Sanctus, Dominus Deus Sabaoth.
Pleni sunt Coeli, et terra gloria tua.
Hosanna in excelsis.
Benedictus qui venit in nomine Domini.
Hosanna in excelsis.

CHAPTER THREE

SECOND SPECIES: THE PASSING NOTE

In this species the counterpoint moves at twice the speed of the cantus firmus.

The note which coincides with the C.F. note is called the "accented note" and the note which falls between the C.F. notes is called the "unaccented note".*

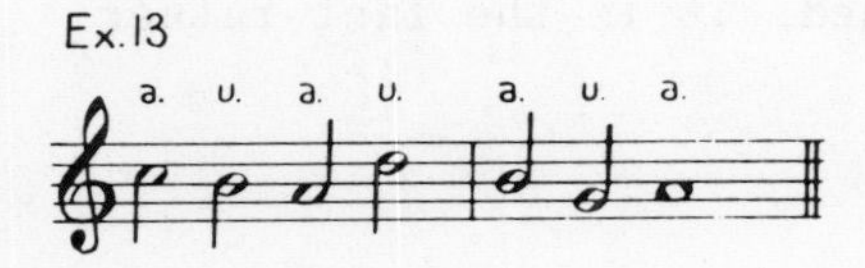

*"Accented note" and "unaccented note" are used in this book as technical terms; thus the second and fourth minims (and the second, fourth, sixth, and eighth crotchets) are described as unaccented notes. Used in this way the terms have an important part to play in the clear understanding of the style. But the musical accent on the words may in fact be different, especially if syncopations are involved. See Ex. 26 and compare Ex. 27 in order to see how flexible the actual word accent is, but remember that this in no way affects the placing of passing notes, etc.

The rule which we now add to those for first species is very simple:

The unaccented note, provided that it moves stepwise between consonances a third apart, may be discordant.

A note making a discord of this type is called a "passing note".

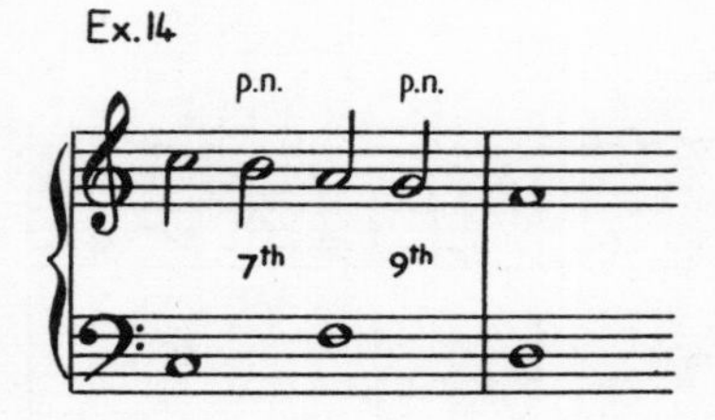

A most important corollary must be understood from this, one that will apply throughout the study of this style of counterpoint:

Leaps may only be made to, or from, concords.

Parallel fifths and octaves

As far as the sixteenth century is concerned, the situation is quite clear, and the rule strictly kept:

Adjacent parallel fifths and octaves are always avoided.

However, as far as fifths are concerned, it is the fact rather than the sound that is objected to:

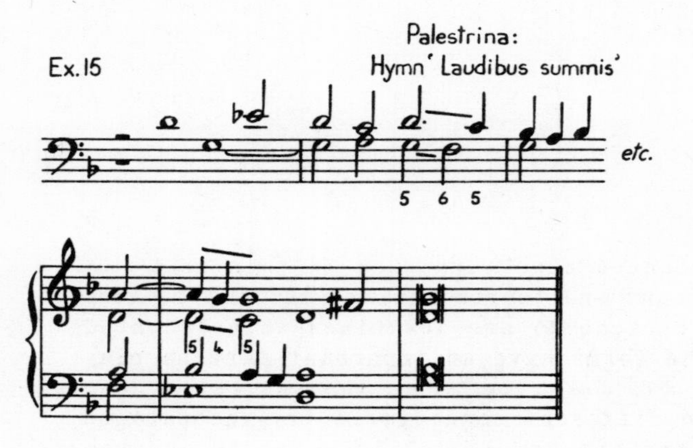

In other words, any consonant or dissonant interval occurring between the pair of fifths renders them innocuous.

It is possible to find hundreds of examples of such narrowly escaped parallel fifths. The search for examples of octaves similarly treated is rewarded by comparatively few specimens. In works for five or more voices a composer may take occasional advantage of such a possibility, but only under pressure. Here is an example from Palestrina's six-part mass "Papae Marcelli", where it occurs in two inside parts, one beginning a phrase, the other ending:

We may sum up the rules thus:

(1) Adjacent parallel fifths and octaves should be consistently avoided.
(2) Parallel fifths with one note intervening are possible.
(3) Parallel octaves should be separated by more than one intervening interval.

Apart from the question of parallel movement, in two part writing the frequent occurrence of the perfect consonances (octaves and fifths), especially on accented beats, will produce a rather bare effect, which although not necessarily wrong, may not be intentional. It is normal practice to begin an exercise with a fifth or octave, and in most cases to end so. Otherwise, these intervals should be approached by means of contrary movement in the parts.

The student should note well that contrary movement will provide the solution to many difficulties. "Shadowing" the cantus firmus is bad technique and produces dull counterpoint, even when the intervals are "correct"; and if the intervals involved are fifths or octaves the result is a thin, bare sound and probably some incorrect parallel movement as well:

In order to save time and to begin the study of the C clefs cantus firmi should in future be written in the alto clef and the bass and treble counterpoints written one on either side, in this way:

(N.B. This is not a piece of three-part counterpoint).

If words are not used it is vital that the student should think in shapely and elegant phrases, and express them clearly by means of phrase-marks.

---

*It is worth remembering that there is nothing musically good or desirable in an incessant flow of notes. The use of an occasional semibreve or of a minim or semibreve rest, where this is musically appropriate, may considerably improve a melodic line.

Ex. 1. Add counterpoints above and below in second species:

Restriction of passing-note discords

While the student will benefit from a study of the second species as described above, the author feels strongly the objection that minim movement in the sixteenth-century style makes very little use of the discordant passing note; see the section *'Harmony'* in Chap. 8 and the section *'Rules of consonance and dissonance'* in Chap. 10.

It will therefore be advisable to practise the second species with the following additional rules:

(1) A majority of unaccented notes should be concordant with the cantus firmus. This must not mean that the counterpoint leaps wildly from place to place; a smooth vocal style is still possible, and Ex. 20 below uses two discords only. The move between the fifth and sixth is clearly of great value and may be studied in four different forms in this example.

(2) Such discordant passing notes as are used should move by step **downward** rather than upward.

If the downward-moving passing note is rare, then the upward-moving passing note can almost be called non-existent. There is no point in using now what we must discard later.

CHAPTER FOUR

FIRST AND SECOND SPECIES MIXED

The art of counterpoint is as much the art of combining rhythms effectively as of putting notes one against the other.

Notice in the first eight bars of Ex. 21 for instance, that the rhythm of one part *complements* that of the other - when the top part moves in semibreves the lower part moves in minims, and vice versa. Used mechanically the result would be deplorably dull, but fortunately there are many ways of combining the basic note-values:

(1) The dotted semibreve followed by a minim (a) is equal to half a bar of first species followed by half a bar of second, and obeys the rule proper to each species in turn.

Ex.22

The point marked * must therefore be consonant since it is technically an accented minim. The minim which follows the dotted semibreve however, is unaccented and may be a passing note. This rhythm is the best way in which faster (minim) movement can "grow" out of slower (semibreve) movement.

(2) The semibreve tied across a bar-line to a minim (b) is of course rhythmically identical with the dotted semibreve, and must also be followed by a minim which may be a passing note. This is an example of a rhythm which shifts the accent from the beginning of the bar, one of the ways in which counterpoints are given independence*. A rhythm of this sort is known as a **syncopation.**

*In the canonic second half of Ex. 21 the lower voice will accent in exactly the same way as the top voice, in spite of the bar-lines.

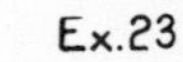

Both these rhythms:  are better if set against a part which moves on the accent; they combine well with each other:

or with plain semibreves or minims:

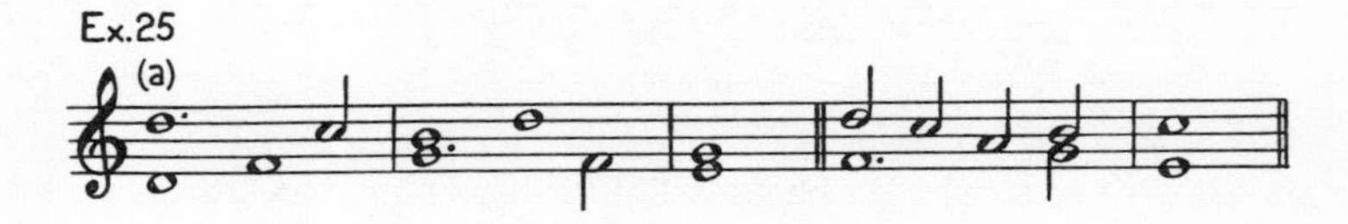

but both parts should not simultaneously avoid the accented beat:

This device may be used only when more experience has been acquired and more resources are available.

The next example shows syncopation of a more striking type:

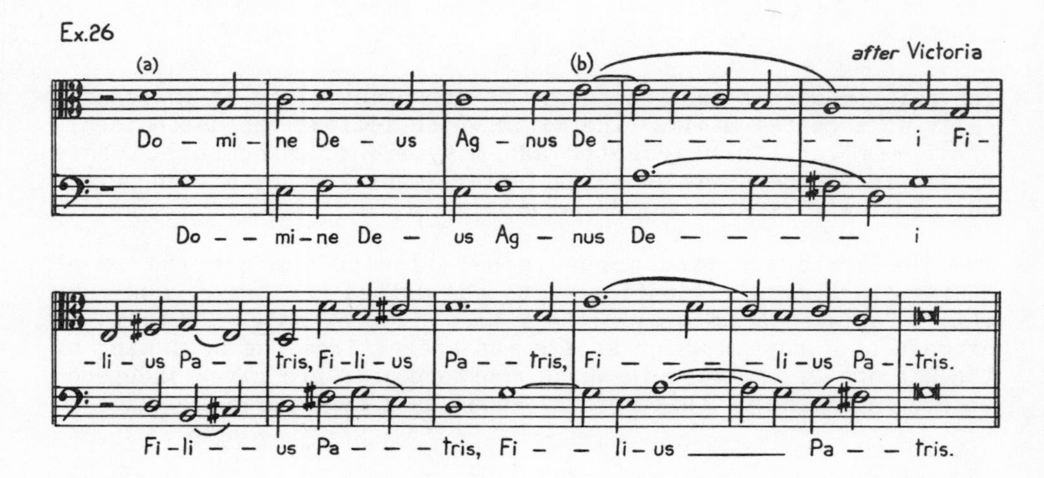

In this case we syncopate minims - that is, we tie an unaccented minim to an accented minim. The unaccented minim then becomes accented - it "robs" the accented minim. The situation is perhaps at its clearest when the note is tied over a bar-line as at (b), but the semibreve at (a) is in exactly the same case.

The result of this syncopation is a very fluid alternation between duple and triple rhythms in the setting of the words, with the bar-lines acting as guides for performance and regulating the placing of dissonances. The following example shows an exact notation of the rhythm:

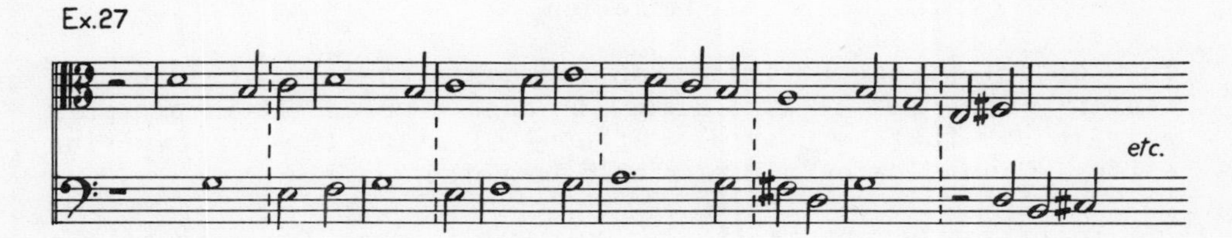

The subject is not dealt with until Chapter 6, but in Ex. 29 the use of dissonant syncopations or **suspensions** may be observed, at cadence points. These are such an invariable part of the sixteenth-century cadence that the student is strongly recommended to use them. The following formulas may be copied and will serve as an introduction to the extensive usage of more advanced work.

Rests

Although it was considered bad at other periods, in sixteenth-century music the moment a part rests it ceases to exist as far as the other parts are concerned. Thus in the second bar of Ex. 29 the treble part is allowed to move to C, hardly a possibility by the theoretical standards of later periods, since the D in the lower part would be considered to persist in the ear long enough to cause an incorrectly resolved discord.

Imitation

The consistent use of imitation is such a feature of sixteenth-century style that it must almost be made a rule. It is probably safe to say that no examples of non-imitative two-part writing exist. The following points should be noted:

(1) Imitation may be at the octave, but is most frequent at the fourth or fifth.

(2) It need not be carried on for more than four or five notes.

(3) Not more than a bar and a half should elapse before the entry of the second voice.

(4) The melodic steps of the theme should be reproduced in the answering voice; i.e. a major third should be imitated as a major third, not as a minor third, and tones and semitones should retain their identity:

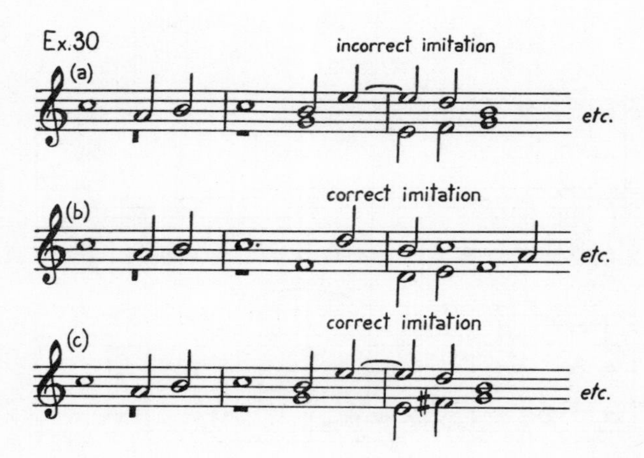

(5) As in the above example (c), an accidental may provide a perfectly acceptable solution to this difficulty. The fourth and seventh degrees of the scale are the only ones that will normally be affected. See also Chapter Ten, p.57

(6) After the main point of imitation at the beginning there may be a number of subsidiary imitations during the course of a piece. In order to introduce these without making unstylish halts and gaps in the flow of sound the phrases are overlapped or "dovetailed". In practice this means that before one part has finished with the final note or notes of its phrase the other part begins on the new subject. Both parts do not stop at the same moment.

Ex. 26 and 29 should be studied with this in mind and note should be taken of the way in which rests are used - partly to give point to the following entry, and partly to give a more punctuated, concise shaping to the melodic line.

Ex. 1*

*When looking for an imitation to a given part, *all* the possibilities should be inspected before deciding which was intended. Don't force an imitation in by the scruff of its neck.

After working the exercises the student may compose some similar pieces. These beginnings may be used.

Ex. 2. Finally, time permitting, a number of canons could be written at all the various possible intervals and distances. Statement (4), p.20 above, can in this case be interpreted with some degree of laxity.

CHAPTER FIVE

THIRD SPECIES: MELODIC DECORATION

Now that we are accustomed to seeing the semibreve and minim used for writing melodies in much the same way that crotchets and quavers are used nowadays, we can learn to handle the faster moving crotchet correctly as an embellishment or melodic decoration.

Since we are dealing with vocal music it should be obvious that the faster the movement the smoother the melodic line must be; rapid leaps cannot be sung either accurately or elegantly.

We distinguish the crotchets as alternately either accented or unaccented:

The rules are similar to those for second species:

(1) The accented crotchets are always consonant with the cantus firmus.

(2) The unaccented crotchets may be consonant; or they may be passing notes of the type used in second species; or they may be *lower auxiliary notes* - i.e. moving from the consonant note one step downward, then back to the original note:

Study the next examples carefully, marking the unaccented note as above:

Melodic movement

(1) After a scale passage, do not leap in the same direction. Only in very few cases will the result be musical. The effect of a "gapped" scale in these circumstances is a very fair musical rendering of the sensation caused by descending a ladder with a missing rung.

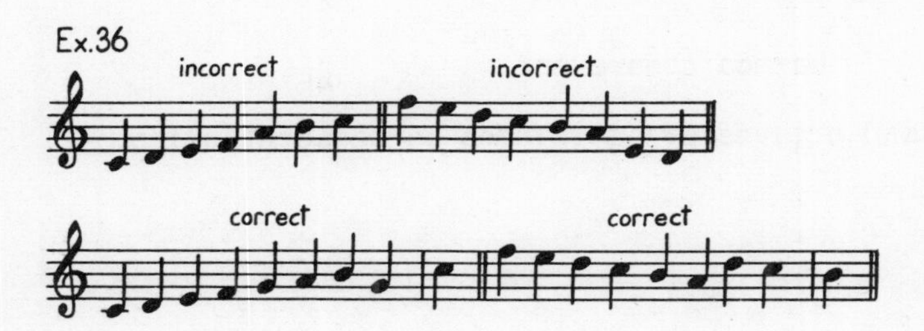

(2) After a leap, move back within the leap. The wider the leap the more important this becomes.

(3) Avoid outlining the tritone.

(4) The leap of the sixth is generally not used. In an emergency the upward leap of a minor sixth may be allowed.

(5) Upward leaps may not be made from accented crotchets.

This last rule is almost invariably kept in third species movement. The student who finds great difficulty in observing it is reminded that third species is less a type of melodic line than a series of decorative patterns, more or less conventional and stylized. It is for this reason much more restricted in movement than the previous species. The correct method of learning it is to study the sample patterns shown in Ex. 44 and then to compare them with the use of crotchet movement in actual sixteenth century compositions (Ex. 112 to 125 for instance). This should demonstrate quite clearly that while some patterns are very popular, others, apparently equally good, are completely neglected or used only very rarely.

Harmonic movement

(1) Parallel fifths and octaves are forbidden on adjacent notes:

(2) Parallel fifths and octaves are better avoided between consecutive *accented* notes:

(3) The interval of the octave should be avoided if it coincides with consecutive notes of the cantus firmus:

(4) Exposed fifths and octaves are still to be avoided:

Ex.43

Keep in mind that the **fourth** and **diminished fifth** are both still as dissonant as ever!

The conventional formulas or idioms shown below are not the only possible moves in third species. The student will naturally invent new moves in order to get out of tight corners. These are the best and most usual in the style, however, and they should form the basis of third species work. Many of them are subject to slight variations; it is the outline that is given.

Ex.44

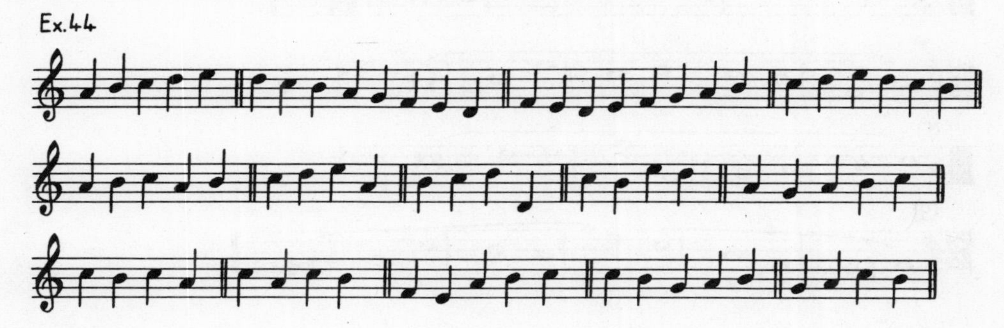

Although academic examination requirements are usually for uninterrupted crotchet movement, musical good sense points out that more convincingly artistic results might result if an occasional note of longer value were used, as in the following example; but this must be done well or not at all, and most certainly not as a way out of a difficulty.

Ex.45

Ex. 1. Write third species counterpoints to the following cantus firmi:

CHAPTER SIX

FOURTH SPECIES: THE DISSONANT SYNCOPATION OR SUSPENSION

In all music there should be a feeling of movement. The composer must be able to generate in the mind of the listener a state of tension which compels the attention and leads in a direct path from the opening bars of the work to the relaxation and sense of completion of the close. In polyphonic music there are several means to this end: the rise and fall of the melodic line, the successive entries of imitating voices, and the use of dissonance.

So far the dissonances that have been considered have all been of a very slight nature. Occurring as they do on unaccented beats and in stepwise movement they hardly disturb the basic consonance of the harmony, and we now consider the more compelling effect of a dissonance falling on the accented beat. The only consistently used dissonance of this type in sixteenth-century music is the suspension. Its characteristics are:

(a) It is a syncopation, and
(b) it "resolves" - i.e. it falls one step to a consonance

At this stage the student should study the use of suspensions in Ex. 146, and in any other pieces to be found in the later part of the book. It will be clearly seen that the dissonance is not actually struck - percussed dissonance is not a

part of the sixteenth-century composer's equipment, except in a very limited sense.

We may now take the following typical example of a suspension and analyse it:

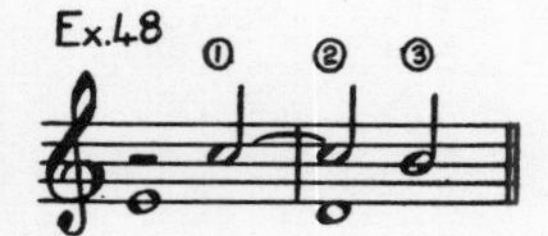

The basic parts of a suspension are:

(1) the **preparation**; a consonant unaccented minim (or occasionally semibreve).
(2) The **suspension** itself;* a dissonant accented minim.
(3) The **resolution**; a consonant unaccented minim.**

The process may be regarded as a delayed move to a consonance, one part falling behind and catching up a minim later. It should be stressed that the preparation must be consonant with the other part, and the suspension must fall one step to the resolution, which should also be consonant.

---

*The three parts together are loosely referred to as the "suspension"; properly speaking only the dissonant accented note should be so called.

** For the sake of accuracy we should point out that other note values can be suspended; minims are normal.

Suspensions above the cantus firmus

In two-part writing there are three dissonances available: the 7th, the 4th, and the 2nd:

Ex.49

Though each has its own character, as suspensions they vary considerably in effectiveness.

Best is the 7th falling in resolution to the 6th, a strong dissonance followed by the full sound of an imperfect consonance:

Ex.50 *

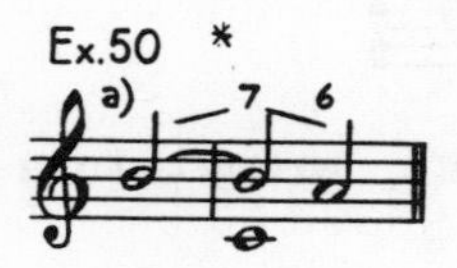

Rather less good is the 4th resolving on the 3rd, for here the dissonance is less pointed, although the resolution is equally effective:

The least usable, the 2nd falling to the unison, is better avoided for the time being. The resolution is ineffectual and the parts are momentarily confused:

* The diagram 7 6 is used in this book to indicate a suspension; the apex representing the dissonance itself, and the two arms respectively the preparation and the resolution.

Some of the objections to its use can be removed by wider spacing, and the 9th falling to the octave is feasible although the resolution is still bare in sound:

In combination with a descending bass the 7:6 and 4:3 suspension will form a "chain" in parallel movement with the base:

This situation must be avoided in the case of the 9:8 suspension since it will result in parallel octaves:

Owing to its resolution on the octave this suspension must in all cases be approached in contrary motion, otherwise exposed or parallel octaves will result:

Its best and most characteristic use is in this converging movement:

Suspension below the cantus firmus

Again there are three dissonances available - the 2nd, 4th, and 7th:

The 2nd resolving to the 3rd is completely satisfactory, for the reasons given above:

Both the other possibilities suffer from the disadvantage of falling to a bare sounding perfect consonance; they will therefore be used, like the 9:8 suspension, only in contrary motion:

It is evident that with only one really reliable and effective suspension available, suspensions in the lower part will be less varied and if good results are to be obtained much will depend on a careful placing of the 2:3. There is, however, the 5:6 move:

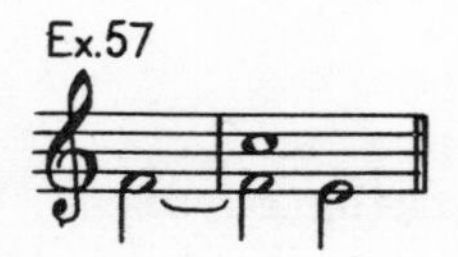

This is not technically discordant in two parts, but it will become, when more parts are used, a good and effective bass suspension, and it may be noted that the ear tends to supply the extra note needed to make it a discord:

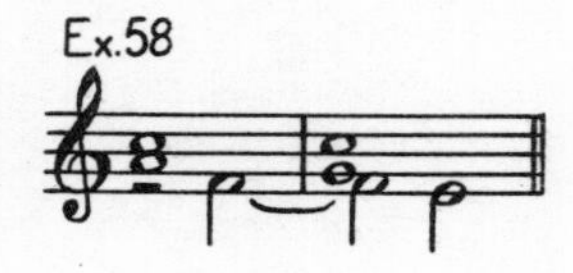

The 2:3 and the 5:6 may be used in descending chains, and in the case of the 5:6 no objection is taken to the fifths on the accented beats:

Ex.59

Ex. 1. Indicate by figures (thus 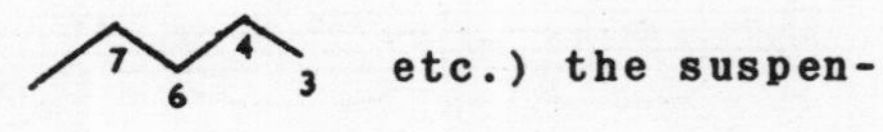 etc.) the suspensions in the following examples:

Ex.60

Ex.61

Ex. 2. Add a counterpoint above, placing a properly prepared and resolved suspension where marked.

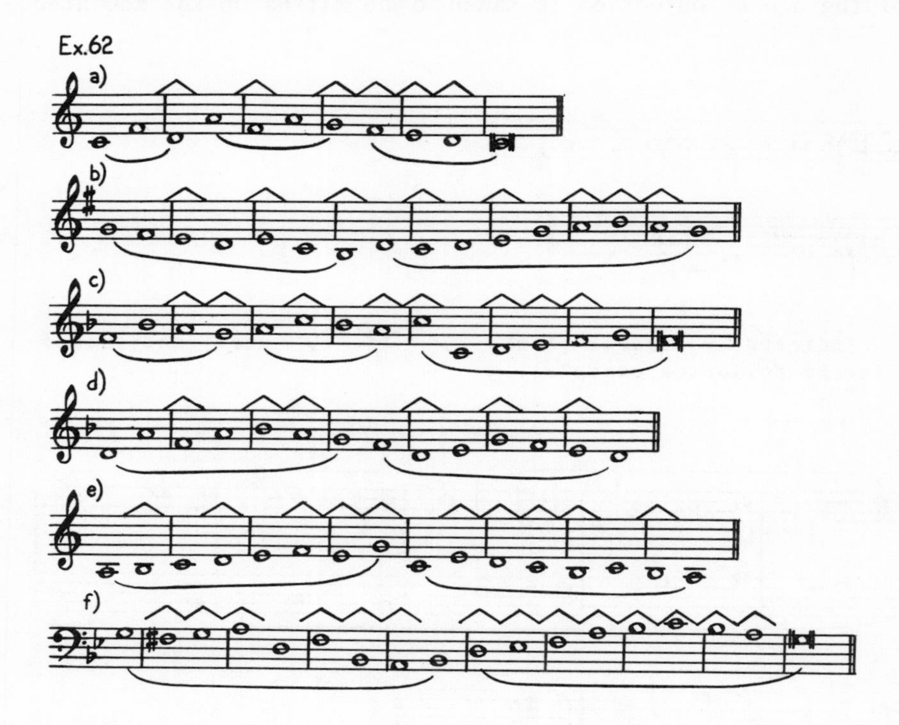

Ex. 3. Add a counterpoint below, placing a properly prepared suspension where marked.

Continuous syncopation

For further practice, and to satisfy academic examination requirements, it is worthwhile practising fourth species in the following way, making use of both consonant and dissonant syncopation to form a continuous chain of syncopations:

Care must be taken not to arrive at an untenable situation. Where this happens the exercise must be corrected by reworking from a point where an alternative move was possible, or from the beginning. If, as occasionally happens, a cantus firmus is quite unworkable, then the syncopation must be interrupted and a bar of second species used to overcome the difficulty.

CHAPTER SEVEN

THE SYNTHESIS: FIFTH SPECIES

In third and fourth species we have studied the decorative elements that give life and colour to the basic melodic shapes of first and second species. Our task now is to combine all the available types of movement into a flexible and well-moulded melodic line that will be both beautiful in itself and also in combination with other similar melodies. At this point, if at no other, author and teacher alike can only say: "such and such are the possibilities - you must learn to choose the right ones with only instinct to guide you", a constantly recurring situation in all forms of artistic work.

It is fortunate that, in such a case, instinct can be trained by making a practical contact with the music of the sixteenth century; those students who know little or nothing of this period should begin at once to make themselves familiar with it. Until it is a living language to them the study of the technique of writing it can only be a dull and profitless exercise in the manipulation of intervals.

But no student who has sung through the masterpieces of sixteenth-century music can make the mistake of thinking it a dry and academic study. I have rarely known anyone who has been involved in a performance of the Palestrina Mass, "Papae Marcelli", for instance, who was not completely convinced of the beauty and nobility of the style. The student will note that I do not say "no one who has heard one of the masterpieces, etc." - it is the experience of singing that is important, not that of hearing. No listener will ever quite understand the singer's thrill in bringing Palestrina's wonderful melodic lines to life as part of a polyphonic web of perfectly organized sound.

One advantage here is that this particular style in the hands of Palestrina, Victoria, Lassus, etc., is in many ways very simple to sing. The restrictions placed on melodic movement that we have already noticed ensure that no awkward intervals will occur, and the problem of sight-singing is very much simpler than in more recent works. Almost the whole of the difficulty for most singers is in reproducing the subtle and sometimes complex rhythms that appear at times. A steady beat from one of the singers,* careful counting of long notes and rests, and a little familiarity will soon dispose of all these troubles. Most important of all, a "good voice" is not needed - often indeed the trained singer is at a disadvantage, since a powerful voice with a vibrato will not blend in a group of singers so satisfactorily as the simple pure tone of an untrained voice. Clear sounds and good intonation are all that is needed.

Evidently therefore the first exercise in this chapter must be:

Sing any available sixteenth-century music. (If nothing else is easily obtained, then the many pieces quoted later in this book will do for a start.)

---

*This beat is known as the "tactus" and was always in use in the sixteenth century. It was traditionally associated with the pulse-rate, giving a tempo around =80 M.M., although there is little doubt that the tempo could vary within the same work. The hand movement is straight up and down

and provides a time-base or reference point for the various "cross-rhythms".

The following four melodies show the typical melodic line of a sixteenth-century composition. This line is a fusion of all the rhythmic possibilities that have been previously dealt with and is academically regarded as a "fifth species".

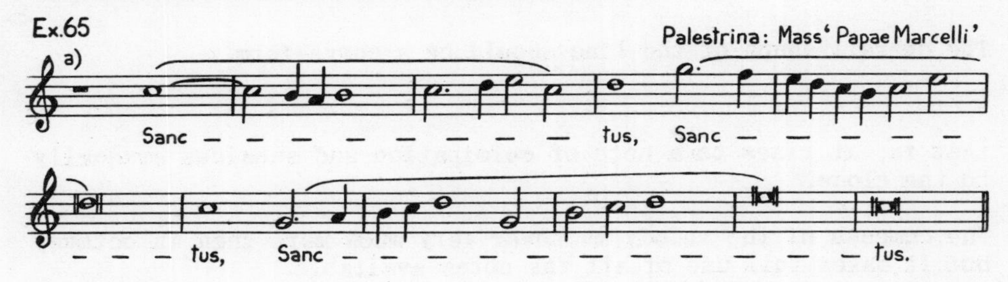

Melodic line

The overall shape of the line should be a curve-form:

that is, it rises to a note of culmination and subsides gradually to the close.

The compass of the melody is never very much more than an octave, but it makes full use of all the notes available.

In other words:

> It is not repetitive,
> It does not hang about the same few notes,
> It does not put undue emphasis on any one note.

The best way of testing a melody is to sing it; if it is at all difficult or awkward there is probably something wrong. Remember that as movement increases so leaps decrease; the faster the smoother.

Finally, the law of compensation; a move upward is compensated by a move downward, and vice versa. A melody never rushes wildly from one register to another; it moves in loops and convolutions as natural as the flight of a bird or the outline of a range of hills.

Rhythm

(1) **There is a perpetual variety of rhythm.** The repetition of rhythmic patterns, which is normal in Bach or Handel, is completely avoided in this style. The following rings quite false:

Ex.67

It is considerably more convincing if rewritten in a way that avoids the use of symmetrical rhythms:

Ex.68

(2) It follows from the above that sequence should be avoided, except in the most fragmentary form.

(3) Furthermore, in an exercise of reasonable length it is unnecessary to use the same rhythmic pattern twice.

(4) Melodies tend to start with long notes and gradually increase in movement; they come to a close with a gradual decrease in movement. But it is not a *symmetrical* increase or decrease - this would be a travesty:

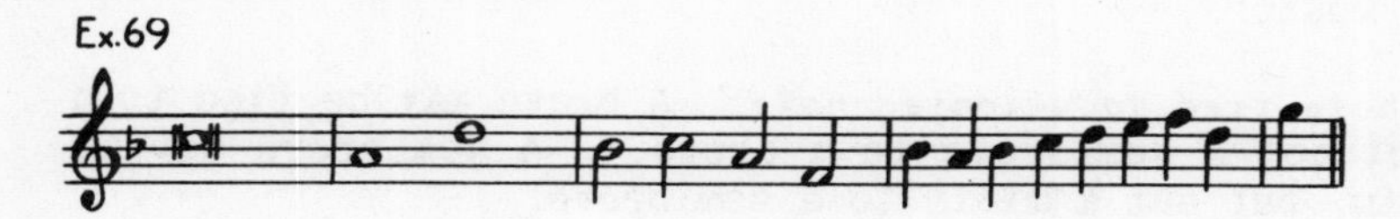

(5) Minims and semibreves are quite often repeated, especially to give an emphatic beginning to a phrase:

This very springy opening may be copied. In general the beginner is advised to beware of overmuch repetition of notes. This may cause moments of stagnation in the melodic line. Exx. 116 and 154 show the correct use of repeated notes.

(6) Crotchets are never repeated except in the idiom known as the anticipation. This is found in nearly all cases in association with the suspension and will be dealt with in the next chapter.

(7) The syncopation is of vital importance. Whether dissonant or consonant it is the most powerful means of avoiding a stagnant melodic line. Once again, sing through the Benedictus (Ex. 65). A composer of the present day might have notated it like this:

This looks clumsy; it is difficult to read at sight; the accents are frequent and will probably be over-emphasized by the performer. Palestrina's notation, when one looks at it again, is refreshing in its simplicity, and the words clearly indicate both the position and the strength of the accents.

The student should take full advantage of this expedient - a syncopation could be used at least once in every two bars, and the resulting change of rhythm should be carefully studied.

These devices are not used in the greater part of sixteenth-century vocal music:

(1) *A short note tied to a longer note.* A breve may be tied to a semibreve, but not a semibreve to a breve.* A semibreve may be tied to a minim, but not a minim to a semibreve.

(2) *Syncopation of crotchets.* A crotchet is too fast a movement to be syncopated or suddenly halted.

(3) *A note tied to a note of less than half its value.* The note values available to us are the breve, the semibreve, the minim, and the crotchet. The only way (short of changing the time signature) in which these could be modified, was by dotting them (single dots only, double dots were not in use).

The student must keep in mind the simple fact that **neither bar-lines nor ties** were used. Therefore, these combinations of note values do not occur.

* Except where the breve is the final note of the piece.

The only combinations of note values in use are those that can be expressed by simple means, without ties or double dotting:

Crotchet movement

The most cursory examination of the examples to be found in this book will show that the majority of crotchet movements begin on an unaccented beat, following a dotted minim or its equivalent:

Refer to Chapter Four, paragraph three, and note that the same applies; this is the best way in which faster movement can "grow" out of slower movement.

Of the other possibilities, it is perhaps wise to be wary of crotchets in pairs. They can produce a stodgy, foursquare rhythm if used frequently. In the place of an unaccented minim, or followed by a syncopation (which robs them of their accent), they may be used in moderation:

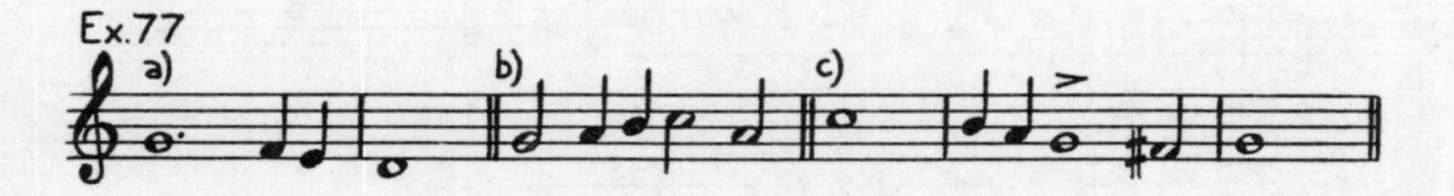

The examination of Ex. 112-125 with an eye to their rhythmical characteristics will be of more assistance than many words of description.

Cadence

The cadence almost invariably makes use of a suspension; as Thomas Morley says, "there is no coming to a close without a discord, and that most commonly a seventh bound in with your sixth as your plainsong descendeth";* i.e:

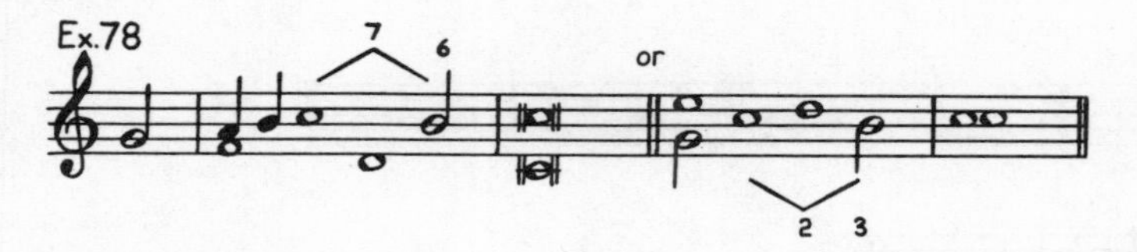

Ex. 1. The following melodies contain many faults, some glaring, some subtle; make a list of them.

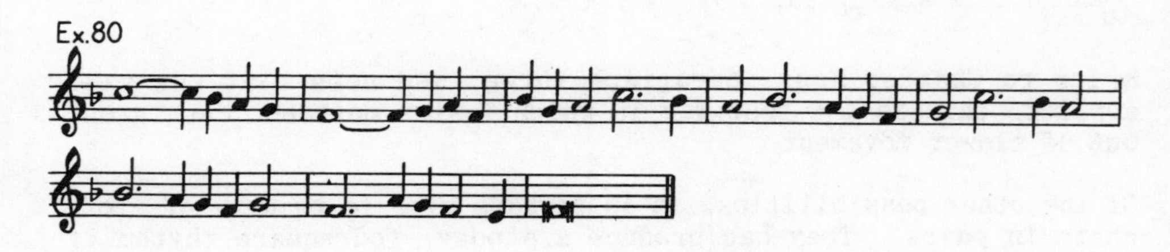

Ex. 2. Write a Kyrie on the lines of the model below.

*A Plain and Easy Introduction to Practical Music (1598)

Ex. 3. Write melodies to the following words. Continue with this exercise until you can construct at will a flowing and expressive melodic line of from eight to twelve bars with not more than two phrases of approximately equal length.

Dies sanctificatus illuxit nobis,
Venite, gentes, et adorate Dominum;
Quia hodie descendit lux magna in terris;
Haec dies quam fecit Dominus:
Exultemus et laetemur in ea.

(Each line should be set to a separate melody)

CHAPTER EIGHT

FIFTH SPECIES WITH A CANTUS FIRMUS

Having considered the melodic and rhythmic aspects of fifth species melody alone, we now learn to combine it with a cantus firmus.

Suspensions

The treatment of suspensions can now be varied by means of the **anticipation.** This can be used either before the preparation:

or as a decoration of the suspended note itself:

but not elsewhere. It is always associated with a suspension. With the anticipation available to soften the dissonance the 2:1 suspension may occur: It obeys the same

rule as the 9:8.

It is important to understand that in the following type of phrase the crotchet marked thus * must on no account be dissonant.

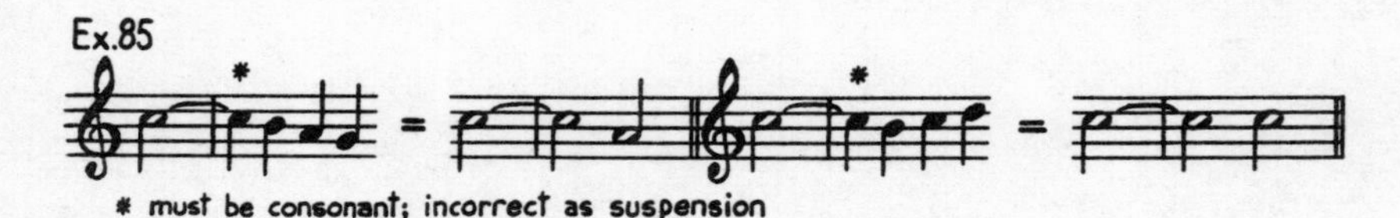

* must be consonant; incorrect as suspension

A correct suspension must resolve on an unaccented minim beat:

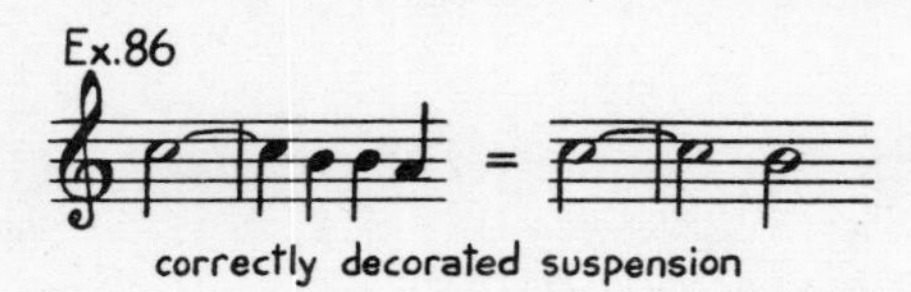

The anticipation does not resolve the dissonance; it merely **anticipates the resolution**. The same applies to the unaccented dotted minim, of course:

Quavers

We now consider the **quaver-group.** The crotchet is already a fast-moving note (equal, perhaps to a semiquaver in modern times), so that quavers will be very fast indeed. If they are to be sung they must always be in stepwise movement. Furthermore, they are always used in groups of two at a time, taking the place of an unaccented crotchet (just as groups of two crotchets preferably take the place of an unaccented minim). Rhythms of this sort, then, are not used:

The possibilities can be summarized quite briefly; quavers following dotted minims:

or, much less good and frequent, quavers following crotchets.

Ex.90

If the condition of stepwise movement is fulfilled, either or both of the quavers may be discordant.

Their most frequent occurrence is in the decoration of cadence suspensions (see Exx. 92c and 124) as a mordent or lower auxiliary on the anticipation. These common figures should be learnt:

Ex.91

In general, quavers should be used sparingly, remembering that they are tricky to sing, and that too many of them will cause a slowing down of the tempo and rhythmic congestion.

Harmony

It is time to revise thoroughly the rules of consonance and dissonance. There are so many possible moves at every turn that no difficulty will be found in observing them. The main rule can be expressed in this simple form:

There will be consonance at every minim beat, unless, (a) there is a suspension on the accented beat, or (b) there is a downward moving minim passing-note on the unaccented beat.

The student should also revise the rules concerning crotchet movement in Chapter Five.

Although the combination of melodies is our first concern, it should be clear that we must not be deaf to the resulting harmony. Particularly at cadences we must feel a strong move toward the dominant, just as we do in conventional four-part harmony. While at other points we are at liberty to use weaker progressions if the part movement is good, ugly sounds are no more tolerable in this style than elsewhere.

Rests

Rests will not normally be necessary in the working of these exercises except at the beginning. That is not to say that the use of them is wrong, if there is a sufficient reason. The smallest value used in the style is the minim rest.

The following models should be studied carefully before beginning work:

Ex. 1. Add counterpoints to these cantus firmi:

Ex.2. Write a cantus firmus to these words and add a counterpoint.

Credo in unum Deum
Patrem omnipotentem,
factorem coeli et terra,
visibilium omnium, et invisibilium.

CHAPTER NINE

THE MODAL SYSTEM IN THE SIXTEENTH CENTURY

The modes that were described in Chapter One gradually evolved, over a period of some five hundred years, into the major and minor scales that we know today as the basis of the music of Western Europe. The pure diatonic modes seem to be completely satisfying only when they are used for unaccompanied melody - plainsong or folk-song, for instance.

As soon as voices are combined together the ear finds certain progressions more satisfactory than others, and in particular a strong cadence is felt to be the most natural way of closing a musical sentence. Although neither of these two progressions can be said to be "better" than the other, certainly the second is more final:

and it was perhaps inevitable that it should be used as a method of punctuation. It involves the use of a leading note (i.e. a note one semitone below the tonic), and since this note is not present in all of the modes, the seventh degree of the scale is sometimes sharpened by means of an accidental.*

The student will realize that in the sixteenth century, scales are at a moment of transition. The modes are used with a good deal of freedom, and although the idea of modulation is a thing of the future, it is possible to make cadences (with sharpened leading notes) on various different degrees in each mode. Usually a piece will move freely between a number of different tonics, and it may be difficult or impossible to establish the mode of the whole piece until the final cadence.

---

*This sharpening of the leading note at cadences was practised quite early on in the history of polyphony. At first forbidden by the ecclesiastical theorists on the grounds that it obscured the distinctive qualities of the individual mode, it was well understood that although the composer conformed to theory and wrote without accidentals, the singer supplied them wherever necessary. It was for this reason that the accidentals were known as "musica ficta" or "musica falsa"; when a piece of music is being transcribed from an early manuscript, care must be taken that all such accidentals are supplied.

We may divide the modes into the following categories:

Major type modes

Ex. 95a

Ionian

The Ionian mode bears the closest resemblance to the diatonic C major scale; however, the free use of accidentals (in particular B flat) make the actual sound of the mode, as opposed to the simple scale, quite distinct.

Ex. 95b

Mixolydian

The Mixolydian mode is similar to the Ionian, since an F sharp will invariably be used at the cadence. But if a B flat is used there may occur an alternation between major and minor tonic chords which is characteristic of the mode.

Ex. 95c

Lydian

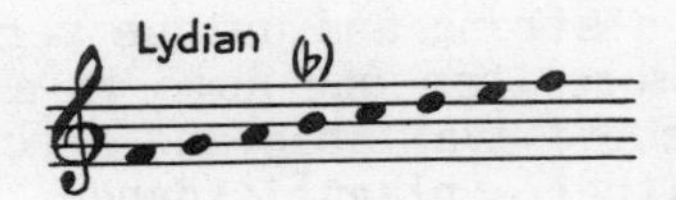

The Lydian mode is so difficult to use, lacking a subdominant chord, that B is invariably flattened, which makes it identical with the Ionian. It may therefore be disregarded.

Minor type modes

Ex. 95d

Dorian

The Dorian mode is one of the most frequently used. It takes a C sharp at cadences, and if as well a B flat is used it has a great similarity with the Aeolian mode.

Ex. 95e

Aeolian

The Aeolian mode takes a G sharp at cadences, and has many characteristics in common with the modern minor scale. It does not normally use B flat.

Ex. 95f

Phrygian

The Phrygian mode has a strong and unique personality; since we do not normally sharpen more than one note in a chord in this style, there is no dominant chord available, and hence no need for a leading note. It makes either a plagal cadence, or this:

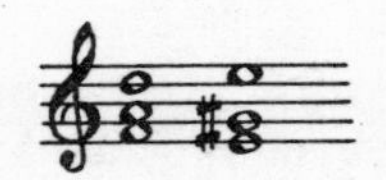 which we may well designate the "Phrygian cadence".

In the course of studying examples of the use of modes it should become clear to the student that it is rather exceptional for a piece to remain strictly in the same mode. For instance, Ex. 126, "Absolve Domine" by Matteo Asola, although it begins and ends firmly in the Dorian mode, makes an excursion into the Ionian almost immediately (half close, bar 4) and after beginning a return (bar 7) makes a very decided cadence in the Ionian at bar 10. In Ex. 161, the Palestrina 'Agnus Dei', a very convincing Ionian opening (bars 1 - 8) is followed by a suggestion of Phrygian (bars 8 - 16), and finally the true Mixolydian mode of the mass as a whole is established in one of Palestrina's simple but wholly satisfying conclusions.

To sum up, then, we have five modes - Dorian, Phrygian, Mixolydian, Aolian, Ionian - and the accidentals which occur in them are C sharp, F sharp, G sharp, and B flat. These accidentals may occur freely in any one mode, forming these cadences:

Broadly speaking, there is a tendency to cadence more often on the dominant than elsewhere, where this is possible (the Aeolian cannot, lacking a supertonic, and the Phrygian cannot, lacking a dominant; they substitute the cadence on the subdominant). The minor-type modes also tend to make fairly frequent use of the mediant close, and the major-type modes of the submediant. Nevertheless, it is the rich variety of possible cadences that makes the modal system so effective an alternative to the diatonic-modulating system of more recent periods, and one should remember that it is in the nature of modal music to make occasionally an unexpected move.

Transposed modes

Some of the examples quoted will be found to have a key signature of one flat. This is a simple device to bring the pitch of the modes into a different relationship with the "tessitura"(or middle compass) of the voice. The modes are then described as "transposed Dorian", "transposed Phrygian", etc. Any further alterations in pitch were made simply by beginning a few semitones higher or lower. The use of this device seems to stem from the identity of Ionian and Lydian.

The final close

One further point must be mentioned. The "Tierce de Picardie" is possible at most closes, and absolutely obligatory at the final close, as the slight discordance of a minor third was considered an unsatisfactory end to a piece. A bare perfect consonance was preferred.

The following examples should be studied and sung or played as illustrations to this chapter:

Dorian: Exx. 92(c), 97, 112, 116, 118, 126, 150
Phrygian: Exx.. 138, 141.
Mixolydian: Exx. 113, 125, 147, 161.
Aeolian: Exx. 115, 117, 124.
Ionian: Exx. 146, 149, 154.

CHAPTER TEN

THE TWO-PART FUGAL STYLE: COUNTERPOINT WITHOUT A CANTUS FIRMUS

The student is now equipped to begin the study of two-part polyphonic composition in the sixteenth-century style without the support of a cantus firmus. First learn, by singing or playing, this two-part motet by Orlando di Lasso:

Amongst the many things we have to learn from it, perhaps the most striking characteristic of this motet is its consistent use of imitating entries, producing an absolute equality of interest in the two parts and that "unity in diversity" which is characteristic of good counterpoint.

Rules of form

(1) When a phrase has been stated and imitated it proceeds to a close; at bar 5, on the subdominant (C), at bar 9, on the dominant (D), at bar 15, on the mediant (B flat), at bar 19 on the tonic (G), at bar 21, on the dominant (D, plagal cadence), and finally, on the tonic (G).

(2) Each new phrase of the text is set to a new musical phrase. The last phrase is repeated three times, building up to a not unimpressive climax by means of faster note values. In this case the thematic material is treated differently each time. At bar 16 the theme (nos tuo vultu saties) is stated; at bar 19 it is inverted; at bar 22 it is imitated in inversion, by which time it has become varied from its original form.

(3) Each entry of a new theme is overlapped by the conclusion of the last close. A glance through the other examples in this chapter will show that we may make a rule that both voices should not rest at the same moment.

Rules of consonance and dissonance

In general, the rules already learnt are still completely valid; however, certain new rhythmic combinations can arise and we will do well to summarize the whole situation. The basic rules are given first; the student will then find a number of exceptional cases, all common turns of phrase which break the basic rule and which are here designated as "idioms" - colloquialisms or fixed habits of speech that do not vary and may break the laws of grammar.*

---

* They are usually survivals from earlier and more primitive periods when some arbitrary dissonance was tolerated. The fifteenth and sixteenth centuries neatly reverse the present trend to ever more dissonant chordal combinations, and the great masters of the period we are dealing with made a final clearance of the inconsistent elements of the harmonic usage of previous schools. For further discussion of this topic, see Knud Jeppesen *Counterpoint* and *The Style of Palestrina and the Dissonance*.

(1) On every minim beat there must be a consonant interval, except where a suspension is used. Take any or all of the examples in this chapter; this point may be demonstrated by running the eye down and noting the interval at each beat:

(2) It follows that the minim passing note is to all intents and purposes forbidden, since not one example of its use can be found. It will reappear in three and four-part work.

(3) In note against note combination - semibreve with semibreve, minim with minim, crotchet with crotchet - the intervals are always consonant.

(4) Unaccented crotchets may be dissonant as passing or auxiliary notes when in combination with any longer note-values.

(5) Quavers can occur on unaccented crotchet beats, and either or both may be dissonant with any other note value.

Imitation

The intervals of the subject are imitated as exactly as possible in the following voice (or "answer"); the step of a tone is imitated as a tone and not as a semitone, the major third as a major third and not a minor third, etc. It follows that the most usual intervals of imitation will be the fifth, fourth, or octave, below or above. Imitation is most strict in an exposition, particularly at the opening of the piece. Subsequent re-entries of a subject can be treated more freely.

The interval of imitation may be determined by the nature of the subject. If the subject moves upward by step, for instance, the easiest answer is "in the fifth"; if downward, "in the fourth"*.

* The answer is said to be "in the fifth" if it is a fifth higher or a fourth lower, i.e., it is always measured **above** the subject.

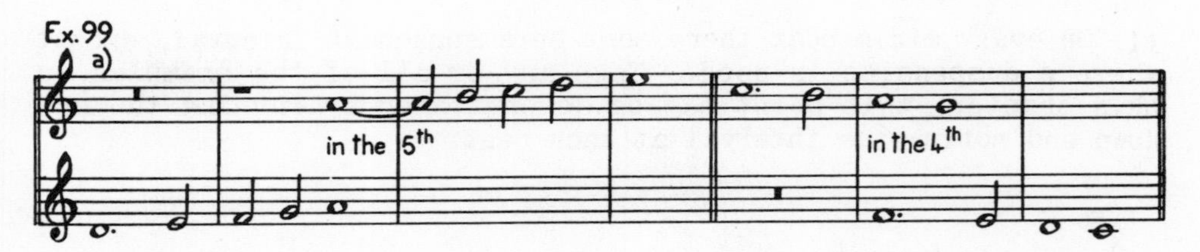

Other intervals of answer are possible if accidentals are used, but this is not desirable, since it will impair the modality.

It must not be thought that slight discrepancies in imitating voices are absolutely forbidden. Different composers have different methods. Palestrina goes to some lengths to maintain a consistently exact series of entries in his mature works, discarding all irregular solutions except under extreme pressure. Victoria, while equally meticulous, uses far more often than Palestrina the "tonal answer", in his constant search for the neatest and most elegant solution of this problem. Where the subject moves between the tonic and dominant of the mode at the outset, this is a great aid to a close imitation (or stretto exposition - see p.102), and also confines the subject and answer firmly to the mode. The rule is:

Where the subject moves between tonic and dominant at the outset, in the answer this move may be reversed.

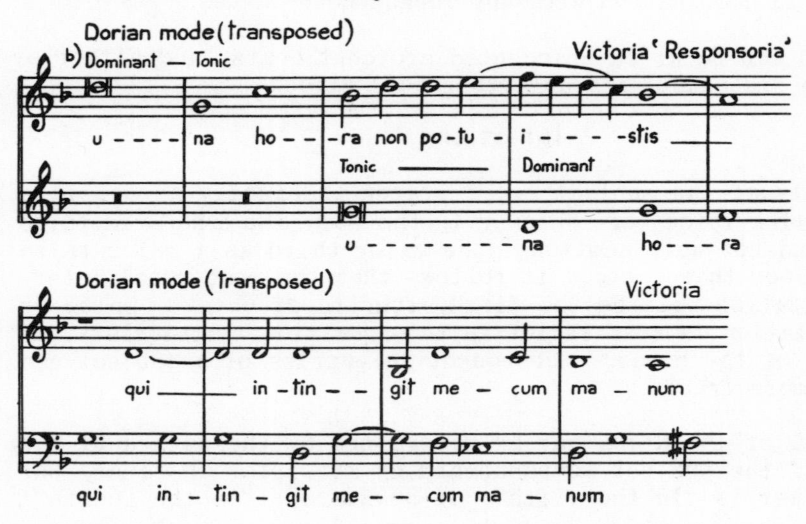

It is probably the feeling for tonality, that is, a desire to define the mode, that prompts this further development that becomes so important in the hands of the Baroque composers:

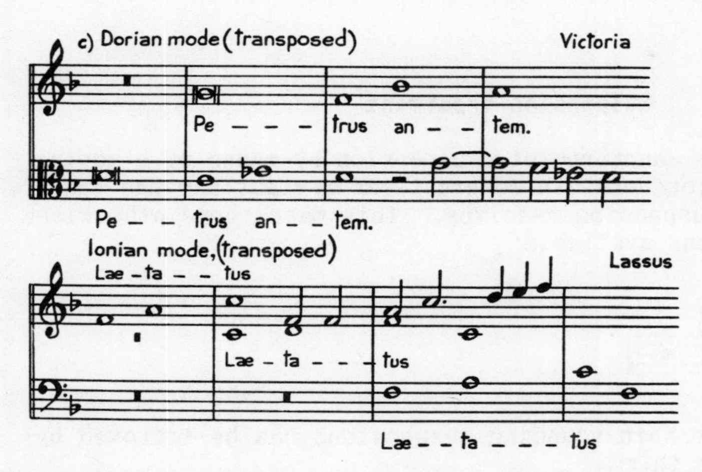

in this case the subject begins on the dominant note, and this is answered by the tonic a fourth higher. The answer then reverts to the fifth. The alteration must be made as soon as possible and in such a way that it does not seriously alter the character of the subject.

Lassus also makes use of this device:

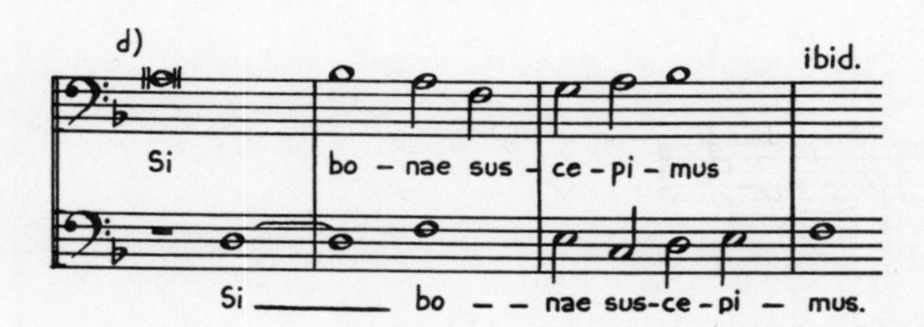

but the student is warned that he is also capable of a completely ad hoc solution such as the following:

Suspension treatment

It is now possible to counterpoint a suspension by means of a moving part in minims or crotchets. In modern terms we might say the chord can change as the suspension resolves. This makes some otherwise impossible suspensions available:

Ex.100

some rather rough or thin-sounding suspensions can be improved by moving to a sixth or third:

Ex.101

and extra movement may be added to normally correct 7:6 or 4:3 suspensions:

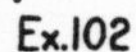

Ex.102

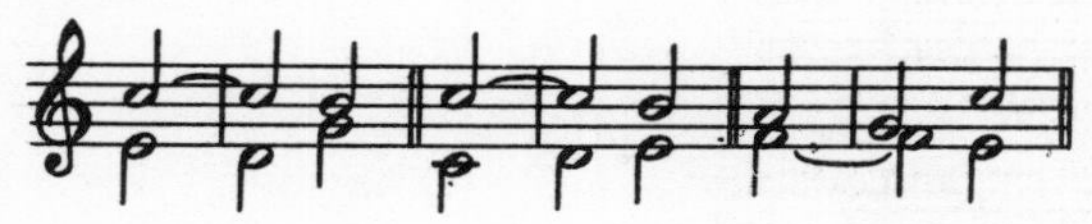

If movement in crotchets is added, then it will in most cases move upward by step or as a lower auxiliary:

Ex.103

The upper auxiliary may also be occasionally used:

Ex.104

Idioms

(1) The "Nota cambiata"* is distinguished by a name to itself - "changed note" - and it is the sole example in common usage of a discord left by leap. The basic shape is:

Ex.105

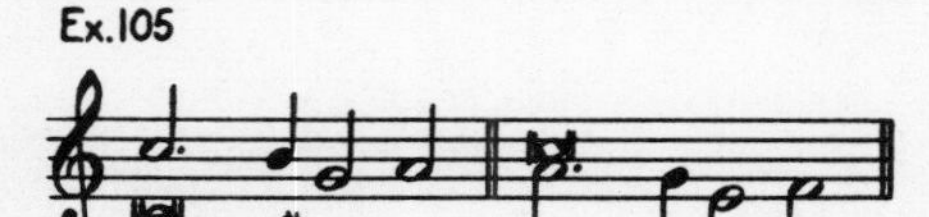

and this shape is always constant. Whatever the context the four notes shown by the bracket remain in that order and no other: a step down, a leap of a third down, a step up. The second note may be dissonant - the first and third must always be consonant.

There are three main rhythmic variants of the cambiata:

Ex.106

these suffice for all normal needs.

(2) The following idiom is more easily illustrated than described:

Ex.107

* The name seems to be incorrectly applied, but is so generally known that it would be confusing to omit it.

This could be described as a two-part idiom, since both the crotchet figure and the suspension occur together and the dissonant accented crotchet marked- x is not found in any other context. Some very minor variants may be found, but the student is advised to use this idiom exactly as given.* Examples of its use will be seen in Ex. 114 and in Ex.118. But note that Lassus clearly prefers a less dissonant version for use in two parts:

Notice this form in Exx. 97, 113, 115, 117, 125.

(3) The most common exception to the first basic rule (p. 57) is this:

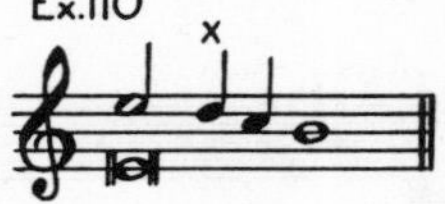

where the crotchet x is dissonant. The explanation of this is that after a longer note the first of a group of shorter notes is felt to be *relatively* unaccented, even though it may fall on an accented beat, *unless* it moves upward - more stress is automatically put on a rising passage. We may state the case in general terms:

The first crotchet descending from a minim or longer note may be dissonant.

The following two examples demonstrate the unaccented character of the dissonance quite clearly:

---

*A possible (if unverifiable) explanation of this idiom is in the undoubted practice of ornamentation of the written notes by the performer. Thus it might take shape as a mordent-type ornament in this way:

One further example may be found in Ex. 118, combined with a rising crotchet scale which adds a good deal to the dissonance. Such a counterpoint would not be found in the more refined composers of the late Renaissance.

Ex. 1 The examples following must be studied carefully, sung or played until completely familiar, then used as exercises in this way:

(a) Copy out the top part.
(b) Reconstruct by adding a bottom part.
(c) Compare with the original.
(d) Correct your working.
(e) Repeat with the other part.

Ex.113
Lassus: 'Magnum opus musicum'
Be – – – – a – – ta, be – – a – ta cu – – – – – – – – – – – – – jus
Be – – – – a – – ta, be – a – ta cu – jus bra – – – – – – – – – –
bra – chi – is sec – li pe – pen – – dit, sec – li pe – pen – dit
– chi – is sec – li pe – pen – dit pre – – – – – – ci – um sec –
pre – – – ci – um.
– li pe – pen – dit pre – – – – – – – – – – – – – ci – – um.
Ex.114
Pierre Certon: Mass 'Adjuva Me'
Do – mi – ne De – us Ag – nus De – – – – – – – – – – – – – – – –
Do – – mi – ne De – – us ag – nus De – – –
– – – – – – – – – – – – – i Fi – – li – us Pa – – –
– – – – – – – – – – – – – – – i Fi – – li – – us Pa – tris, Fi –
– tris.
– – – – li – – us Pa – – – – – – – – – – – – – – – – – – tris.

Ex.115
Lassus
ibid.
Au – di – tu – i me – – – – – – o da – – – – – – – – – –
Au – di – tu – i me – – o da – – – – – – – – bis gau – – – – – di –
– bis gau – – – – – di – am et lae – – ti – – – – – – – – ti – – am.
– am et lae – – – – ti – ti – – am.
Ex.116
Lassus
m.o.m.
In – – – – tel – lec – – – tum ti – bi da –
In – tel – lec – tum ti – – bi da – – – – –
– – – bo et in – stru – am te, et in – stru – am te in
– – – bo et in – stru – am te et in – stru – am te in vi – – –
vi – – a hac qua gra – du e – – ris hoc qua gra – du e – –
– a hoc qua gra – du e – – – ris, hoc qua gra – du e – – – – – – – – –

-ris fir-ma - -bo fir-ma- -bo su-per te o - cu-los me- - os,
-ris fir-ma-bo fir-ma- -bo su-per te o - cu-los
o - cu-los me- - os, o - cu-los me - - - - - - - - - - os.
me- - os, o - cu-los me - os, o - - cu-los me - - - - - os.
Ex.117
Lassus
ibid.
Pu - - - -tru-e - - - runt et cor-rup - - - -
Pu - - - tru - e - runt et cor-rup - - - - tae sunt
- - - tae sunt ci - ca-tri - - ces me - - - - - - - - - - - ae
ci-ca-tri - - - - - - ces me - - - - - - - - - - - - - - - ae
a fa - ci-e in spi-ti - en - ti - ae me - - - - - - - -
a fa - ci - e in spi-ti - en - ti-ae me - - - - ae in spi - ti -
- ae.
-en - - ti-ae me - - - - - - - - - - - - - - - - - ae.

Ex. 2. Exx. 119 and 120 are "unseens", and additional exercises of this kind may be prepared from the collection of two-part motets in Lasso's *Magnum opus musicum*.

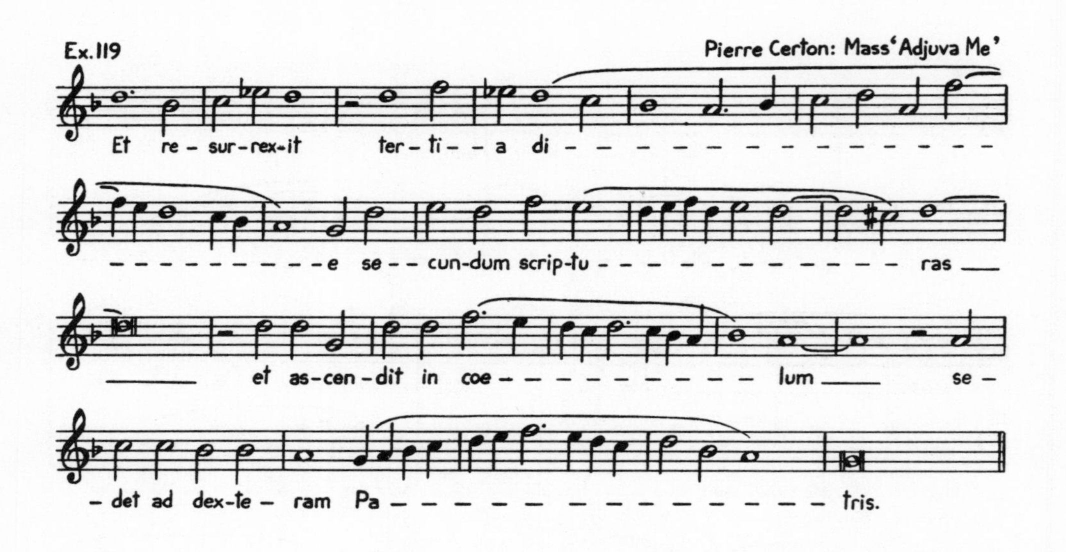

Ex. 3. When facility has been gained in adding parts correctly these words should be set, using the original thematic material of Lassus as given, or original themes.

(1) Te deprecamur largius,
Nostris adange sensibus
Nescire prorsus omnia
Corruptionis vulnera.

(2) Esurientes implevit
bonis: et divites dimisit
inanes.

(3) Justus cor suum tradet
Ad vigilandum diluculo
Ad Dominum qui fecit illum
Et in conspectu altissimi deprecabitur.

Ex. 4. Finally the use of free canon is shown in Exx. 124 and 125. Analyse these examples carefully.

The writing of canon is strongly recommended as a means of strengthening contrapuntal technique. Phrases should be short, lively, and not repetitive. Take advantage of rests, within reason. Avoid symmetrical phrases (the oft-quoted finale of the Cesár Franck violin sonata is a very bad canon from our present viewpoint). It is an easy matter to write a canon that rambles on until rigor mortis sets in; a good and interesting canon requires concentration and practice.

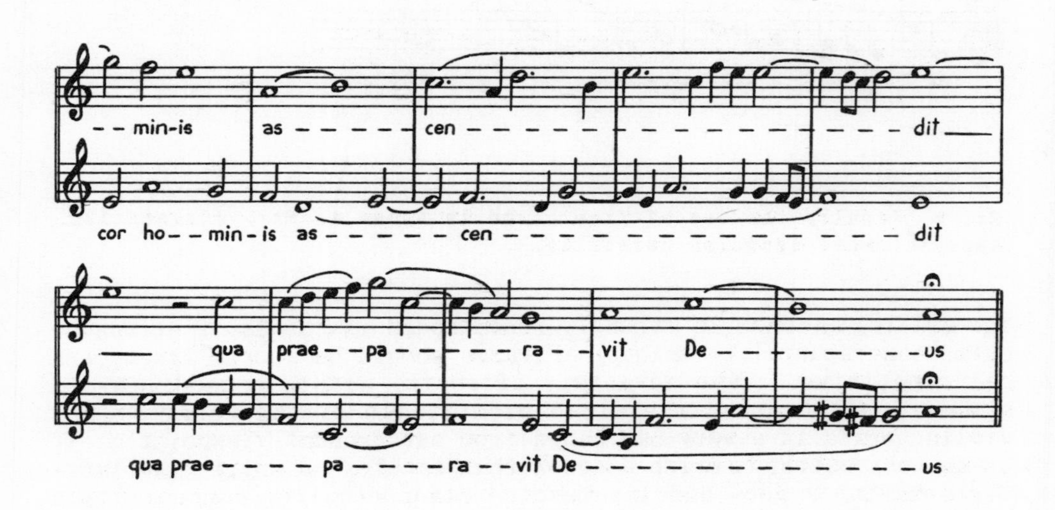
– – min-is as – – – – – cen – dit
cor ho – – min-is as – – – – – – cen – – – – – – – – – – – – – – – – – dit
qua prae – – pa – – – – – – ra – – vit De – – – – – – – – – us
qua prae – – – – pa – – – – ra – – vit De – – – – – – – – – – – – – – – – us

Ex.125
Lassus
m.o.m.
non con-se-quen – – – – – tur, non con-se-quen – – – – – – – – – – – – tur, non
non con-se – – quen – – – tur non con-se – quen – – – – – – – tur
con-se-quen – tur.
non con-se – quen – tur.

CHAPTER ELEVEN

THE HOMOPHONIC STYLE IN THREE PARTS

The first requirement is the ability to hear music in three parts. Although this will gradually develop as fluency is gained in writing in three parts, much can be done to hasten the process. Both silent score-reading and playing from score at the piano or organ should be studied now if they have not already been taken in hand. The second can be studied in class or tutorial, the first only in privacy and without help. If the student has a score always in his pocket for the odd moment, he will learn to read it as a child learns to read a book; that is, by picking out familiar signs and gradually piecing them together, and so in time he will possess one of the most useful and enjoyable gifts a musician can have.

In the sixteenth century there are two textures available; we shall call them "fugal" or "imitative", and "homophonic". The first is self-explanatory and most of our time will naturally be taken up with this style, but our understanding of sixteenth-century music will be very one-sided if we do not consider the important role played by music of a more harmonic character - with parts moving, as it were, in first species, note against note.

Ex. 126

Matteo Asola: Mass 'Pro Defunctis'

Homophonic music of this type has several advantages over fugal music, in certain circumstances:

(a) The words will be heard more clearly;
(b) A larger number of words can be set in a short time;
(c) The effect will be simpler, hence more dignified, more animated, and so forth, depending on the character of the words;

and, of course, it will provide a strong contrast to the fugal style.

Ex. 1 Take a complete mass setting; distinguish between the two textures, and note the effect of the words on style and texture. Write a concise description of each section.

Although masses vary in detail the following is the usual procedure:

Kyrie-Christe-Kyrie...... Fugal
Gloria.................... Mainly homophonic
Credo..................... Mainly homophonic; always homophonic at moments of great solemnity, e.g. "Et incarnatus est"
Sanctus................... Fugal
Hosanna................... Homophonic
Benedictus................ Fugal; often for fewer voices
Hosanna................... Homophonic; may be a repeat of previous Hosanna
Agnus Dei 1 .... Fugal
Agnus Dei 2 .... Fugal; often adding an extra part which may be in canon.

The combinations of intervals that are available to us would be classified nowadays as root position or first inversion triads. These are the basic harmonic ingredients of sixteenth-century music, and all other chords are the result of suspensions or other forms of dissonance treatment. This concept of the inversion of chords would have been quite strange to a sixteenth-century musician, however simple and obvious it may seem to present-day students,* and we must therefore state the situation in sixteenth-century terms:

(1) Intervals are measured from the lowest note sounding, which we term the bass.

(2) Above the bass there may always be the interval of the third, major or minor.

(3) In addition to the third there may be above the bass the interval of the fifth or sixth.

Ex. 127

Normally this way of thinking of chords in terms of intervals produces quite ordinary results.. But it is possible to find exceptional combinations of intervals - major third with minor or augmented sixth, for instance:

Ex. 128

*It was first mentioned, though probably not invented, by Rameau in his *Traité de l'harmonie* (1722), the first modern harmony textbook.

sometimes used by composers for dramatic or pictorial reasons.* We may discard them for the moment (with the exception of the diminished fifth triad which we discuss in the next chapter) and make use of only these resources:

Ex. 129

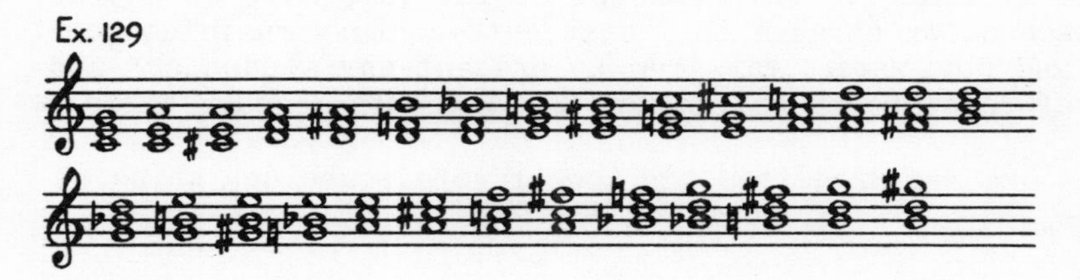

Although it should be clear from the examples, the student should note that the intervals of the augmented fourth and diminished fifth are considered to be concordant when they occur between two upper parts:

Ex. 130

There is no objection to consecutive fourths; they occur freely in this characteristic progression often known as "faux bourdon";

Ex. 131

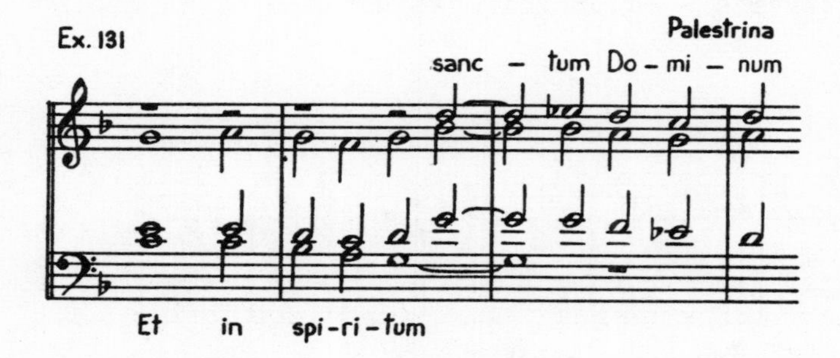

Adjacent parallel fifths are incorrect even if one of them is diminished:

Ex. 132

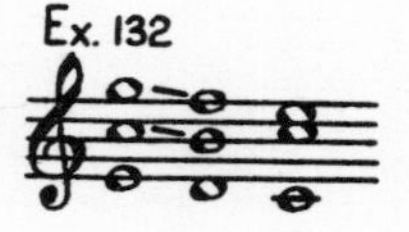

*This does not occur much in the Italian composers we are chiefly concerned with. For a splendid example, see Orlando Gibbons's five-part madrigal "The silver swan".

Part movement

(1) Spacing should be as even as possible without wide gaps between any of the parts.

(2) The maximum number of complete triads should be used, although there is no real objection to a chord with either third or fifth missing.

(3) In circumstances where a complete triad is not possible the part movement usually decides which note is to be doubled; there are no binding rules on the subject, and the third may be doubled as freely as the fifth or root. (But note the next paragraph.)

(4) Leading notes at cadences must on no account be doubled. Neither may any sharpened notes (musica ficta) be doubled, since they always act as leading notes. Flattened notes are free of this prohibition.

(5) Parts may cross if such crossing in any way improves the melodic lines. In the case of a composition for equal voices crossing is normal and expected.

N.B. Take care that a part in alto or tenor clef does not cross **unintentionally** below the bass and produce discordant fourths.

(6) All parts should not move strongly in the same direction. Nine times out of ten this will produce parallel or exposed fifths or octaves.

(7) Conversely, the unsatisfactory sound of an exposed interval can be considerably mitigated by a strong move in a contrary direction by the remaining part:

Ex. 133

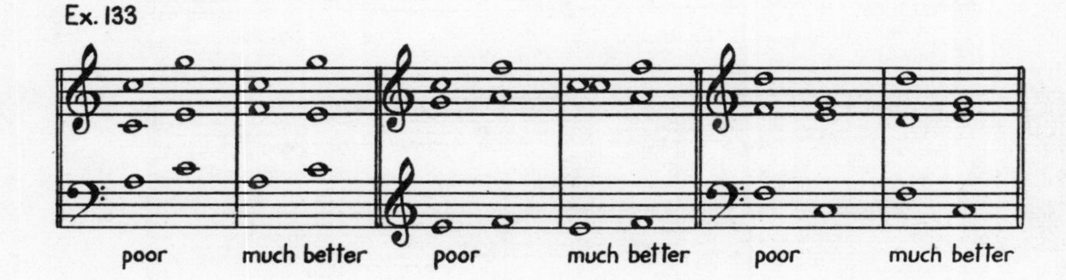

It should be clear that although in this chapter we restrict ourselves to movement in block chords, with all parts more or less rhythmically identical, there must still be a feeling of melodic life in each part. Nevertheless, the lowest part will now take on a more "bass-like" character. Notice in Ex. 126 how the bass moves in strides to strong fifth-related root-position chords.

The treble, since it is most audible to the listener, will naturally receive the greatest care and be fluently and elegantly shaped. It will move less boldly than the bass, making more use of conjunct movement. In general, it will tend to be in contrary motion with the bass.

The inside part, though it will unavoidably receive secondary attention, must never be allowed to stagnate or be in any way awkward or unmelodious. It may occasionally assert itself at a cadence by means of a suspension, as in Ex. 126.

It must be stressed that three-part writing calls for discipline and mental concentration. **It is absolutely useless to add parts one at a time**; the three parts must be mentally conceived, at least in short stretches (two or three chords) at a time, before being notated. Only by thinking in three parts can the conflicting requirements of harmony and melodic line be reconciled. The impatient student is warned that a new and better technique is often slower in its early stages.

Ex. 2. Add two first-species parts to a plainsong cantus firmus, in the style of this example:

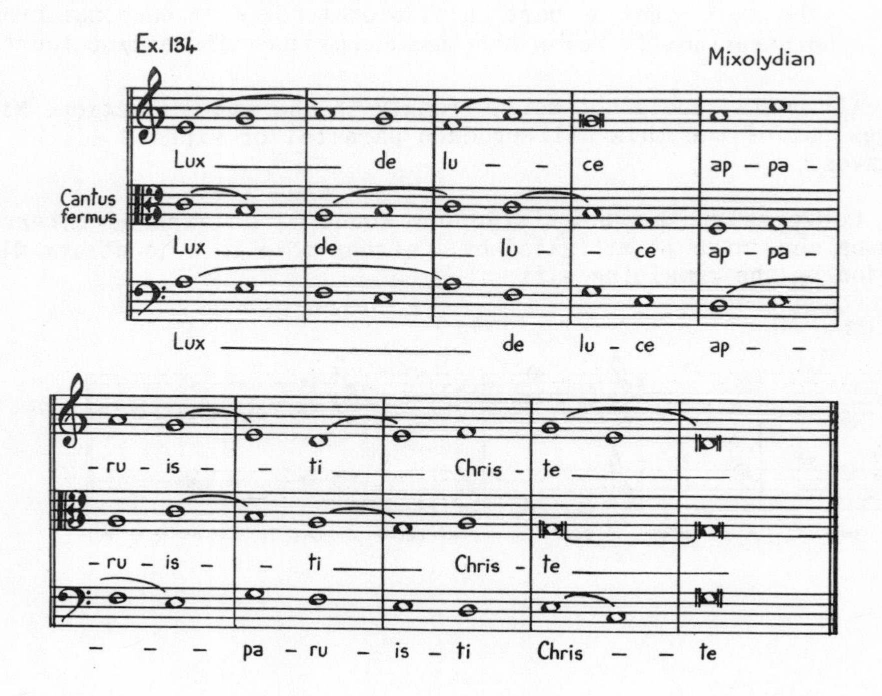

(Plainsong cantus firmi will be found in Appendix 4).

Ex. 3. Add two lower parts:

Ex. 4. Set these words in homophonic style; Ex. 126 should be used as a model. Set each verse to a separate melody.

Stabat Mater dolorosa,
Juxta crucem lacrimosa,
Dum pendabat Filius.

Cujus animam gementem,
Contristatem et dolentem,
Pertransivit gladius.

O quam tristis et afficta,
Fuit illa benedicta
Mater Unigeniti!

Triple Rhythm

A three-in-a-bar time signature occurs occasionally in sixteenth-century music. It is appropriate that it should be dealt with here since it is usually associated with a homophonic texture.

The relationship between the various duple and triple time signatures is an exact mathematical proportion in all cases, although, of course, the *tempo* of the music may vary according to the mood of the moment and the style of the particular piece. It would be impossible here to begin to explain the system of *proportion* as it was in use in the polyphonic period, and, complicated as it is, it is further complicated by the fact that it was, like the modal system, slowly evolving into our own method of notation, and that it was frequently misunderstood by sixteenth-century musicians.

The following statements should cover the situation and bear the test of practical application as far as is necessary at the moment.

(1) The triple rhythm is a faster movement than the duple.

(2) The proportion is - one bar of 4/2 equals two bars of 3/1.

In the case of a work which changes from one to the other, (see Ex. 150 or the well known "Hodie Christus natus est" of Sweelinck,) we must learn to feel this relationship in the way a pianist learns "twos against threes"; it is of the greatest help that the downbeat of the "tactus" (see p. 36) remains constant as a time base.

The student must be warned that many modern editions distort and obscure the relationship of duple and triple times by shortening the note values - presumably in an effort to make the music look more up-to-date.

Ex. 5. Complete the following:

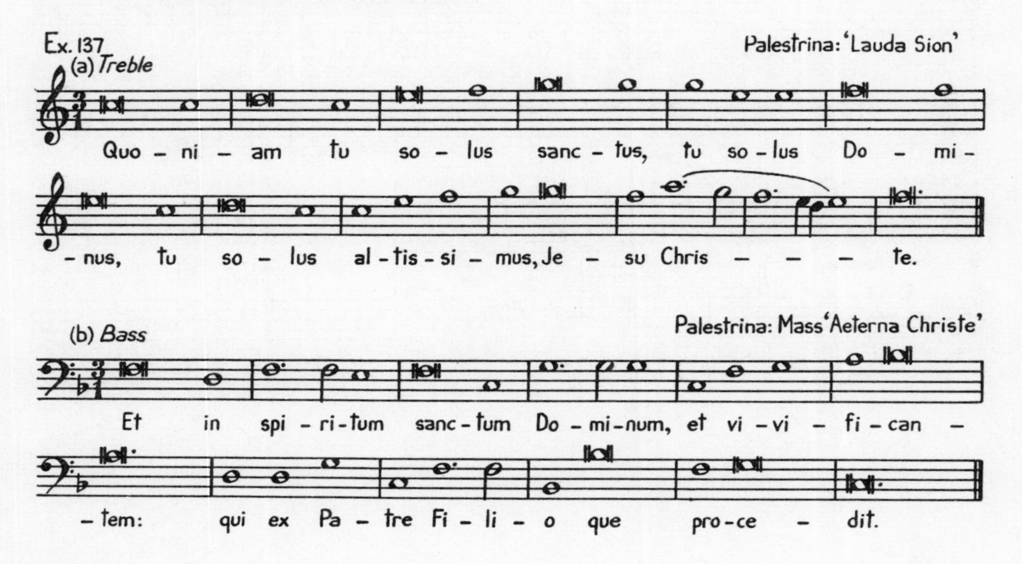

CHAPTER TWELVE

FANTASIA ON A PLAINSONG CANTUS FIRMUS

Before work is started on the task of combining three equal parts, a useful and not unrewarding discipline is to be found in the addition of two contrapuntal, imitative voices to a plainsong. Although we use it as a means of further expanding our technique we must not make the mistake of regarding it as a purely academic routine. It is one of the fundamental procedures in composition of all periods, and is still valid as part of a composer's technique. The principle is the same as that of the Chorale Prelude - the embellishment of a melody, played slowly, by means of faster-moving counterpoints that contrast with it (even though they may be derived by inversion, diminution, etc.) Notice in the example that follows: (a) the counterpoints begin each phrase before the cantus, (b) they make use of diminution of the first notes of the cantus, and (c) in two cases (bar 5 and bar 15) a new imitation is begun during a phrase.

15
Sa – – lus et lan – –
-lus et lan – – – – guor om – – ni – um, sa-
Sa-lus et lan – – – guor om – – ni – um, sa-lus et
– – guor om – – ni – um.
-lus et lan-guor om – – ni – um. Sa-na – –
lan – guor om – – – – – ni – um. Sa –
20
Sa – na – te ae – – – gros
– te ae – gros mo – – – ri – bus, sa – na-te ae-
– na – te ae-gros mo – – ri – bus, sa – na-te ae-gros
25
mo – ri – bus, nos – – red – – den-tes
-gros mo – ri – bus, nos red-den-tes vir – tu – – – – ti – bus, nos
mo – – ri – bus, nos red-den – tes vir-tu – – – – – – ti – bus, nos
30
vir – tu – – – – – – – – ti – bus.
red-den-tes vir – tu – – – – – – – ti – bus.
red-den-tes vir – tu – – – – – – – – ti – bus.

In Chapter Ten the use of the unaccented minim passing note was mentioned and firmly discouraged on the grounds that it hardly ever occurs in two-part composition. It may now make an occasional re-appearance on these conditions:

(1) It must fall to a consonance
(2) It must not produce a percussed dissonance: or, in other words, the other parts must be either stationary, or consonant with the passing note, thus:

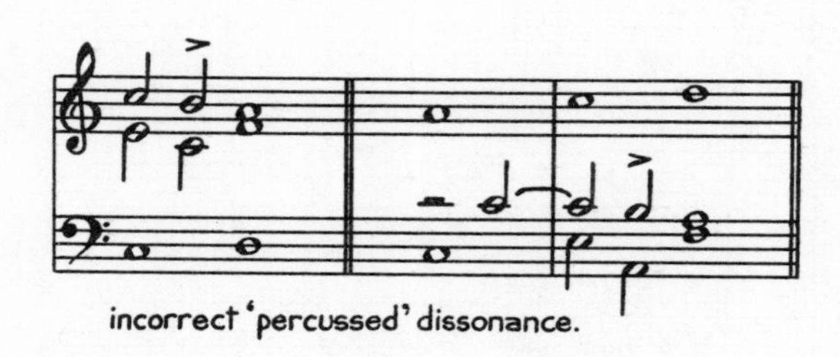

There will, however, be few occasions when the preceding minim cannot be dotted, so that the dissonance is only the length of a crotchet.

Attention should be given to the dissonant combinations of intervals at the beginning of bars 7 and 18 in Ex. 138. Such progressions were popular with the composers of this period, and added greatly to the harmonic resources available. In these days we should define the chords produced as various positions of diatonic sevenths; in earlier times they would be considered as dissonant suspensions with the other parts moving to and from any other available notes consonant with the bass. Provided that the parts were all correct with each other, and that the suspension resolved according to rule, many interestingly dissonant combinations could arise, especially in compositions for many voices.

In Ex. 138 notice also several cases of suspensions which are not tied across the accent but repeated.

The diminished fifth may now occur, provided that it moves conjunctly inward (as it would if it were part of a dominant seventh). In most cases this occurs as the resolution of a discord, as at the first and second beats of bar 18, Ex. 138. Clearly, it is easier to tolerate a slight dissonance if it is immediately preceded or followed by a stronger one. It may also be found in this form:

Ex. 140

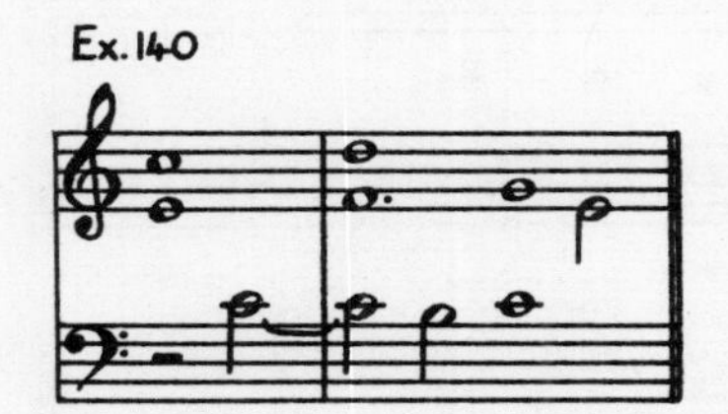

The rarest of all forms of the diminished fifth chord is to be found in Ex. 138. bar 30.

Ex. 1. Write plainsong fantasias on cantus firmi from Appendix Four. The cantus firmus may appear in either the upper or the middle voice, or it may move, phrase by phrase, between the three.

CHAPTER THIRTEEN

THREE-PART FUGAL STYLE

There is much to be learnt from a study of the neat and elegant counterpoint of the next example:

Ex. 141 Victoria: Mass 'Quarti toni'

Notice:

(1) The mainly conjunct, or stepwise, movement of the parts.

(2) That the basic minim movement is not allowed to become monotonous but receives constantly varying accentuation; demonstrating very clearly the importance of word placing in polyphonic music.

(3) The use of imitation. Very few notes are used simply to "fill out".

The 'Consonant Fourth'

At the last beat of bar 7 in the example above there will be found an exception to the rule that the preparation must be consonant. In most cases the "consonant fourth" appears at cadences, thus:

Ex. 142

There are three conditions governing its use:

(1) The bass must be stationary during all three parts of the suspension.

(2) There must be a third part which makes a 7:6 or 2:3 suspension with the part taking the consonant fourth, thus forcing it to its resolution by making a stronger dissonance.

(3) The consonant fourth must be approached by step.*

---

*See also Ex. 126, where the upper part "borrows" the dissonance, moving down and up in an auxiliary pattern not often found in second species movement.

Double suspensions

It is now possible to suspend two notes simultaneously:

In thirds and sixths:

In fourths (usually with one resolution decorated):

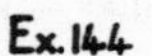

Even in fifths:

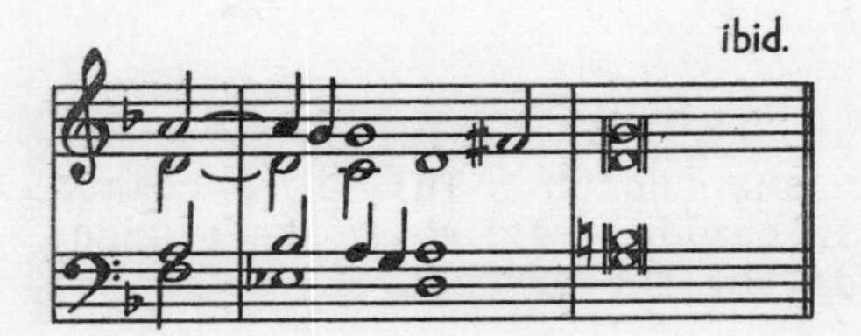

The suspensions in Ex. 141, 146 and 147 should be carefully studied. In particular the use of suspension in the next example is a model of the beautiful effect that can be produced by this means.

In bar 12 there is an example of a rare type of resolution decorated by a note falling a third from the suspension. This is a common figure in the English school, and in earlier periods. The student may consider it as outside his needs for the present.

The next example, Ex. 147, is in a more florid style. The words "qui venit" are set to what can only be described as a **counter-subject**, or second theme. This will be considered further in the next chapter. Note that it forms a **stretto** at bar 10, and again at bar 14 in a decorated version. While the student will find it difficult to emulate Victoria's economy of material it should be noted that here, as in the other examples in the chapter, there is very little "padding". Themes of contrapuntal character often combine in more than one way, and a careful analysis should be made with this point in mind.

While one cannot make rules to cover all the problems that will arise in working the exercises, a few generalizations may help:

(1) Be economical with notes; three-part counterpoint is not one half as complex again as two-part. The law of diminishing returns is at work: **the more parts, the less elaboration.**

(2) Basic melodic movement is in minims and semibreves; crotchet movement is more effective if used with discretion.

The use of a fairly high proportion of crotchets may be seen in the energetic last section of Ex. 150 (vitam que nostram dirige); this may be regarded as a maximum, to be equalled only when the occasion really demands it.

(3) As a corollary to the above: do not be afraid to write long notes.

(4) Try to write long supple phrases, but be content to bring them to a close before they die of exhaustion. The average length of a single phrase is from six to eight bars, after which a rest of a bar or a bar-and-a-half might be made.

(5) Study the examples. And remember that the essence of study is memorization; not parrot fashion, but memorization of the sound and the means by which it is attained.

Ex.1. Copy one of the parts of Exx. 141, 146, 147. Reconstruct and compare with the original. If possible, repeat the process with the other parts until the piece has been virtually memorized.

Ex. 2. Complete the following skeletons. If the original source of the compositions is available it should be compared.

(c)
Palestrina: Mass 'Veni sponsa Christi'
Altus
Tenor
Bassus
Ple – ni sunt (etc.)
Ple – ni sunt coe – li et ter – – ra
Ple – –
– ni sunt (etc.)
Ple – ni sunt
Ple – ni sunt coe – li et ter – – – ra.
(d)
Palestrina: Mass 'Sine Nomine II'
Cantus
Altus
Tenor
Sanc – tus, sanc – – – – – – – – tus, sanc – –
– – – – – – – – tus, sanc – – – tus

Palestrina: Mass 'Sine nomine'
(Je suis déshéritée)
(e)
Cantus
Altus
Tenor
Be – ne – dic – tus, be – ne-dic-tus qui ___ ve – nit, be-ne-dic-
-tus qui ___ ve – – – – – – – nit, be-
– ne-dic–tus qui ___ ve – nit, qui ___ ve – nit.
be – ne-dic–tus

Palestrina: Missa 'Brevis'
(f)
Cantus
Altus
Tenor
In no-mi-ne
In no-mi-ne ___
In no-mi-ne ___ Do –

in no-mi-ne, in no-mi-ne
Do - mi - ni
In no-mi-ne
Do-mi-ni, Do-mi-ni in no-mi-ne
In no-mi-ne
In
In no-mi-ne
Do - mi - ni.
no-mi-ne
ni.
In no-mi-ne
ni.

(g) Add 2nd Treble and Alto
Palestrina: Mass 'Lauda Sion'
Cantus
Be - ne-dic - tus qui ve - nit, qui ve
nit, Be-ne-dic-tus qui ve - nit.

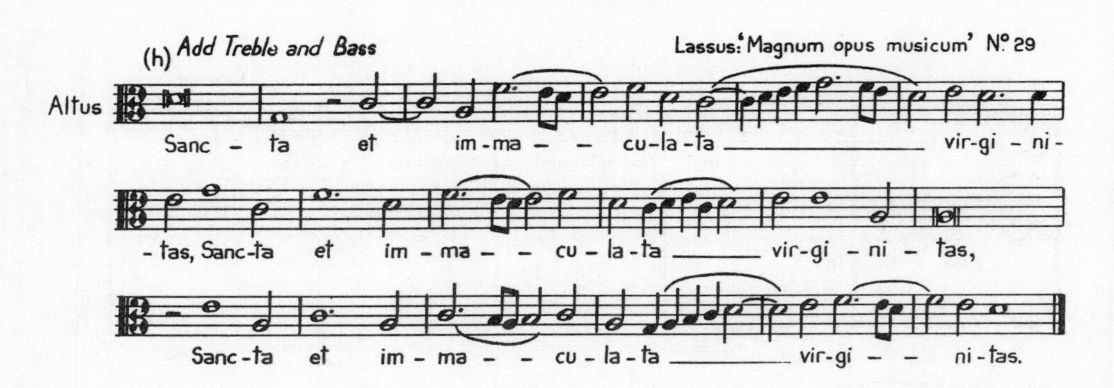

CHAPTER FOURTEEN

THE THREE-PART MOTET

The following section of Palestrina's mass "Regina Coeli" should be analysed. It will be found to consist of a series of "fugal expositions"; with one exception, each new phrase of the text is set to a new theme.

– – – – – – – – bis ____ sub
no – – – – – – – bis sub Pon-ti – o Pi – – – la – –
e – ti-am pro no – bis ____ sub Pon – ti – o ____
10
Pon-ti-o ____ Pi-la – – – to, ____ pas – sus, et se-
– – to, pas – sus, et se-pul – tus est. ____
____ Pi-la – – to, ____ pas – sus, et se – pul-tus est,
15
-pul-tus est. ____ Et re-sur-rex-it ter – – – ti-
Et re – sur-rex-it ter – – – ti – a di-
et se-pul – tus est. Et ____ re-sur-rex-
20
-a ____ di – e, se – cun-dum Scriptu – ras, ____ se-cun-dum
– – – e, se – cun – dum Scriptu-ras, se – – – cun –
– it ter-ti-a di – e, se – cun-dum Scrip-

Earlier in this book a good deal of stress was laid on the proper construction of a melodic line, and on the importance of the rise and fall of tension, in the attempt to keep the listener always waiting for the next move until the finally satisfying close. Now, it is clearly not always easy to reconcile the formal demands of a series of fugues with the demand for a continuously beautiful and well moulded melody. Yet this is in fact what we must strive to achieve, and what the sixteenth-century composers succeeded in achieving. In some cases melodic requirements are given priority, in others the fugal aspect seems to dominate, but in all cases short of pure homophony the two must coexist. The Palestrina "Crucifixus" above is in as purely fugal style as possible, but one has only to sing through the tenor part to realize that it is more than simple filling between treble and bass, and that it has its own life, independent of the part it plays in the fugal texture of the whole piece.

The next example is a typical motet, with a contrasting middle section. In it, Lassus finds a slightly different balance between the two rival claims. The emphasis here is laid on melody, with the top line predominating slightly. But the beautiful melodic sweep of the treble part does not prevent the form from being fugal, and the imitation is in most cases surprisingly exact. The treble voice is simply "primus inter pares" - first amongst equals.

25
-cit ma — trem si — bi, ma — — trem si — — — bi quam e-
-cit, ma — — — trem si — — bi, ma — — — trem si — — bi quam e - le -
-cit ma — trem si — bi, ma — trem si — bi quam ——— e — le —
30
-le — — — — git. ————— Ad - es — — to jam sup — - pli - ci -
— — — — — — — — git. ———— Ad - es - — to jam sup - pli — - ci -
— — — — — — — — git. Ad - es - — to jam sup - pli — - ci -
35
- bus tu — - is fa — ven — - do pre - ci — - bus
- bus tu — is fa — — ven — - do pre - ci — — bus ma — — — num be -
- bus tu — - is fa — ven — - do pre - ci — — bus ma — — — num be - nig-
40
ma — — num be - nig — — — — — nam por - ri -
- nig — — — nam, ma — — num be — nig — — nam por — ri -
— — — — — — nam, ma — — num be - nig — — — — nam por - ri —

-ge vi - - - -
-ge vi - - - - - - - - tam que nos - - tram
-ge vi - - - - - - tam que
45
- - - tam que nos - - - tram vi - - - -
di - - - ri - ge, vi - - - - -
nos - tram di - ri - ge, vi - - - - - -
50
- - - tam que nos - tram di - - - ri - ge, vi -
- - - tam que nos - - tram di - - ri
- - - - tam que nos - - - - tram di - ri - ge, vi -
55
- - tam que nos - tram di - - ri - ge.
-ge vi - - - - tam que nos - - - tram di - ri - ge.
- - - - tam que nos - - - tram di - ri - ge.

Countersubject exposition and stretto exposition

The term "countersubject" is used to describe the counterpoint that accompanies and contrasts with the "answer" (the entry of the subject in the second part). This countersubject, however, not only contrasts with the *answer*, it is also a melodic continuation of the *subject*, and in some cases it will be impossible to decide at what point the subject actually finishes.

Expositions that make use of such a contrasting motive may be termed "countersubject expositions". Their occurrence will depend to a certain extent on whether the words fall naturally into two sections. Clear examples of the technique of countersubject exposition will be found in:

	Subject	Countersubject
Ex. 97	Ipsa te cogat	pietas
Ex. 114	Dominus Deus	Agnus Dei
Ex. 118	Pleni sunt coeli	Pleni sunt coeli (triple rhythm)
Ex. 147	Benedictus	qui venit

A search should be made for other examples.

Another possibility is best described as "stretto exposition". In this case the answer enters so closely on the heels of the first voice that the subject is not complete before the second voice enters. In a composition for more than two voices, a common arrangement is that the first two voices enter in stretto, the third voice being delayed for up to five or even six bars.* It is rarer to find three voices or more in stretto.

Examples of stretto exposition:

Two voices: Exx. 26, 112, 113, 115.

Two voices, third voice entering later: Exx. 141, 149, 150.

Three voices: Exx. 146, 138, 168.

Again, a search should be made for other examples.

---

*In a surprisingly large number of cases the entry of this third voice coincides with a suspension in the other two parts. Both the delay and the suspension enhance the effect of the third entry, and the delay makes it possible for the two parts to move into position for the suspension.

Triple rhythm

An example of the normal style in triple rhythm will be found at bar 28 of Ex. 150; most often homophonic, there is an occasional imitative passage. The following features should be noted:

(1) Breves in 3/1 behave harmonically as semibreves in 4/2
Semibreves " " " " minims " "
Minims " " " " crotchets " "
Crotchets " " " " quavers " "

(2) Thus the melodic line is carried by breves and semibreves; the breve is never dissonant; the semibreve is either consonant or a suspension dissonance, and very rarely a passing dissonance; the minim is allowed either passing or auxiliary dissonance, and crotchets make only rare appearances, in paired groups, etc.*

(3) Suspensions are prepared on the first beat of the bar, struck on the second, and resolved on the third. Exceptions occur at times, and in particular passages of this sort appear to break all rules:

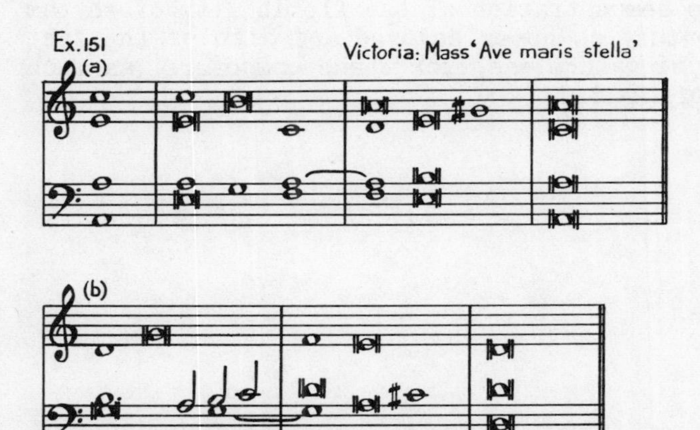

*Owing to the comparative rarity of triple-rhythm time signatures, the characteristic crotchet idioms are difficult to find. As this example shows, there is no reason to doubt that they are possible:

As soon as these are sung the problem disappears: the rhythm has simply moved back into duple time without changing the time signature - it is written across the bar-lines. If we re-bar and halve the note values the passages immediately conform:

Victoria would have been in a position to see this more clearly since he wrote without barlines (or at least with movable ones), and it is an effective demonstration of the flexibility of rhythm that the sixteenth-century composer enjoyed and also of the fact that a change of time signature was, for these composers, as much a change of tempo as of rhythm.

Instrumental music

Much of the music we have to consider is vocal music; certainly the sixteenth-century composers put all their best efforts into composing for voices. But there were other mediums available; many pieces were considered "apt for voices or viols", and there are transcriptions of both sacred and secular music for organ or virginals.

In particular mention must be made of the "canzone", "fantasia", and "ricercare", the instrumental counterparts of the vocal motet and the predecessor of the instrumental fugue. Much of the finest early chamber music is by composers of the English school, Byrd, Gibbons, Tye, amongst many others, and it is hardly possible to do it justice here, since, although the basic principles remain constant, there are many divergences in detail from the practice of the Italian school with which we are concerned.

In this Fantasia by Bassano (a minor but not undistinguished contemporary of Palestrina) notice the purposefulness and logic that is given to the form by the use of thematic material derived in its entirety from the opening subject (bars 1 - 4) and also the recapitulation of this subject at the end. Such formal devices are necessary as soon as there are no words to carry the music on, and we can see here the germination of a seed that has flowered differently in every century. No composer seems able to avoid completely this kind of metamorphasis of thematic material, consciously or unconsciously.

(The student may be amused to notice the subject which evolves at bar 28 - it has received a certain amount of attention from other composers, in most cases inducing in the texture a similar contrapuntal exuberance.)

15
20
25
(piano)
30

35
40
45
(forte)
50
55

Phrase-marks have been added throughout in order to make the thematic development clearer, as well as to help in performance. The phrasing does not indicate a legato performance, although a fairly smooth style seems appropriate. The lack of dynamics does not mean that the piece is to be played without them, but that they are left to the discretion of the instrumentalist, together with the ornamentation.*

Ex. 1. Complete the following skeletons. Compare with originals if available. Further exercises of this type, if needed, may be prepared from the section of three-part motets in the *Magnum opus musicum* of Orlando di Lasso, or the occasional three-part sections of masses by Palestrina or Victoria, etc.

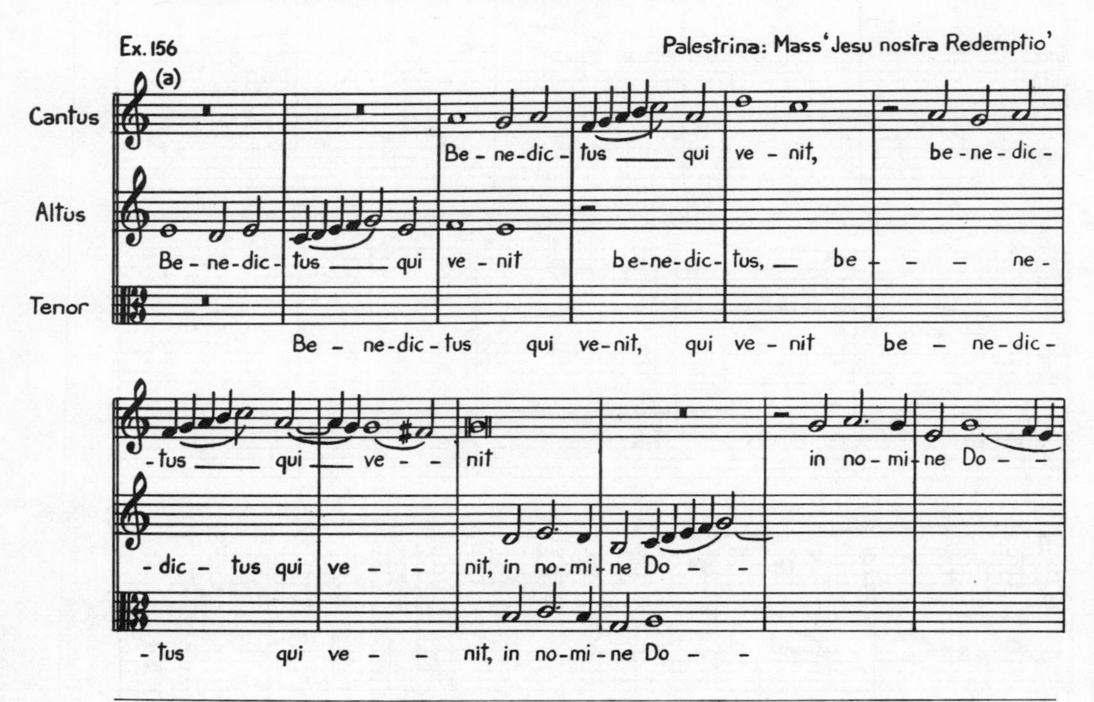

*This is a special study; but the cadences at least should be trilled:

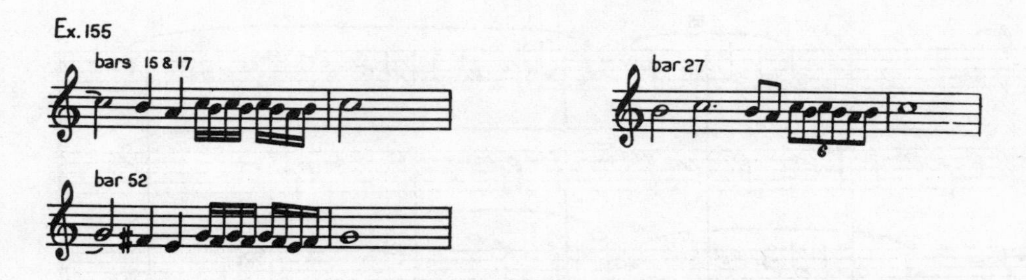

– mi-ni, Do – – – – – – – – – mi – ni, in no – mi-ne Do –
– mi-ni, in no-mi – ne Do – – – mi – ni
– mi-ni, in no-mi-ne Do – – – – –
– mi-ni, in no-mi – ne Do – – – – – – mi – ni.
in no-mi – ne Do – – – – – – mi – ni.
– mi-ne, in no-mi-ne Do – – – – – – mi – ni.

(b)
Palestrina: Mass 'Emendemus'
Altus
Tenor
Bassus
Be – – ne-dic – tus ___ qui
Be – – ne dic – – tus qui ve – – nit, qui ___
Be – ne-dic-tus ___ qui ve – – – – – – – nit, qui ve –
ve – – – nit, be – ne-dic – tus ___ qui ve –
ve – – nit, qui ___ ve-nit, be – – ne-dic –
– – – nit, qui ___ ve – – – nit, ___ qui ve-nit ___

– – – nit, qui ve – nit, qui ve – nit
– tus qui ve – – – – – – nit, qui ve – – – – nit in
be – ne – dic – tus qui ve – – nit in
in no – mi – ne Do – mi – ni, Do – – mi – ni, in no – mi – ne Do –
– no – mi – ne Do – – mi – ni, in no – mi – ne Do – mi – ni
no – mi – ne Do – – mi – ni, in no – mi – ne Do – – mi – ni, in no – mi –
– – – – – – – mi – ni, in no – mi – ne Do – – – – – – – – – – mi – ni.
in no – mi – ne Do – mi – ni, in no – mi – ne Do – – – – – – – – – mi – ni.
– ne Do – – mi – ni, in no – mi – ne Do – – – – mi – ni.

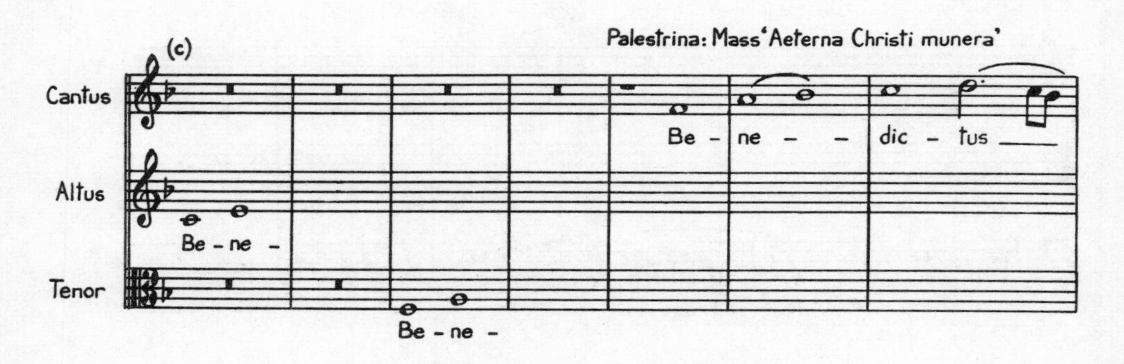
(c)
Palestrina: Mass 'Aeterna Christi munera'
Cantus
Be – ne – – dic – tus
Altus
Be – ne –
Tenor
Be – ne –

— qui ve — — — — — — — — — — nit.
be-ne — — dic-tus qui ve — — — — nit,
in no-mi-ne Do-mi-ni, Do —
— mi-ni, in no-mi-ne Do — — — — — mi - ni.
(d) Dorian mode
Palestrina: Mass 'Spem in alium'
Cantus
Altus
Bassus
Et as-cen — dit in coe — — — — lum, —
— et as-cen — dit in coe — — lum, in coe — —
— — lum, se-det ad dex-ter-am Pa — — — — —
-tris, se - det ad dex-ter-am Pa — — — tris.
Et it — er-um ven-tu-rus est cum
glo — — — — — ri - a,
ju - di - ca - re
vi — vos, et mor — tu — os, vi vos et mor — tu -
- os: cu — jus reg — ni non e — — rit fi —

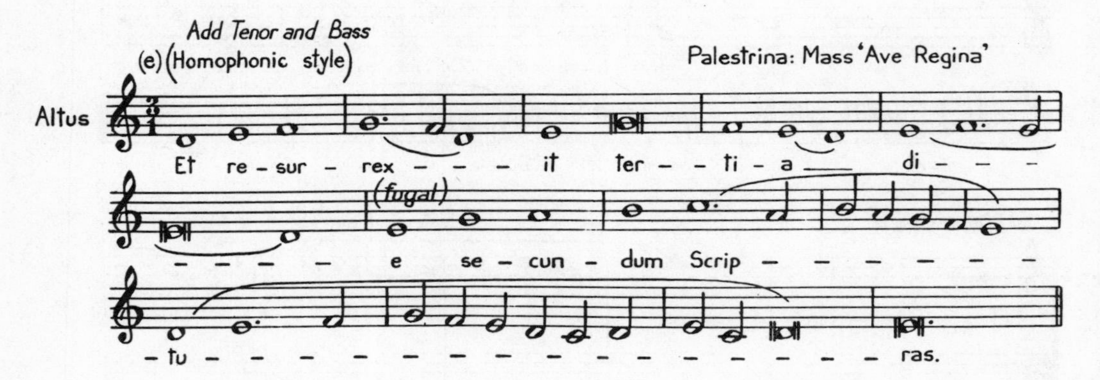

Ex. 2. Write countersubject expositions of the following themes:

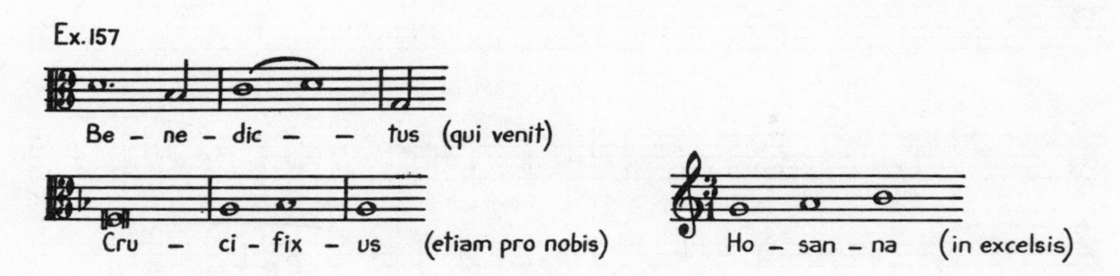

Ex. 3. Write stretto expositions of the following themes:

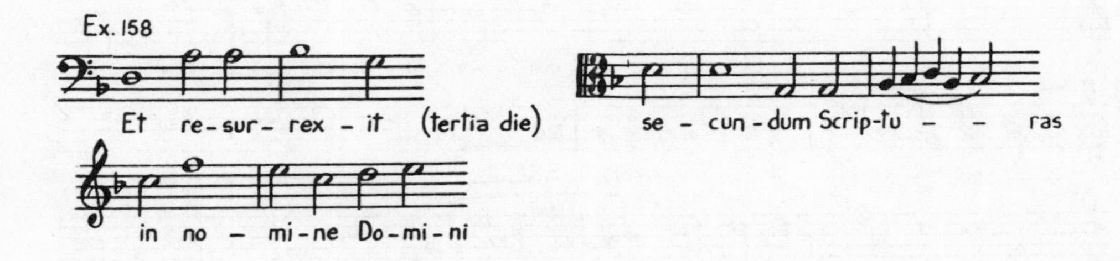

Ex. 4. Make a bar-by-bar analysis of the "Crucifixus" from the mass "Regina Coeli" of Palestrina (149), and of the motet "O maria, Clausus hortus" of Lasso (150).

Ex. 5. Set the following words as a short motet, using the given themes:

Ave Maria,
gratia plena,
Dominus tecum,
Benedicta tu, in mulieribus,
Et benedictus fructus ventris tui Jesus.
Sancta Maria, Regina coeli,
dulcis et pie, O Mater Dei.

Ex. 6. Analyse the Bassano Fantasia (154), making a table showing how the themes derive from the opening subject by inversion, retrogression, and retrograde inversion.

Ex. 7. Using your own material or that given below, compose a short fantasia for three instruments.

CHAPTER FIFTEEN

FOUR-PART COUNTERPOINT

Counterpoint in four parts presents at once additional difficulties and at the same time the means for overcoming them. We are no longer compelled to keep all the voices in constant employment in order to achieve a reasonable sonority, and in most cases there is little more complication in a four-voice motet than in one for three voices. The total effect will naturally be fuller and more impressive but the parts will tend to take longer and more frequent rests and be generally a little less active; a voice will enter with its relevant thematic material and then drop out to make room for another voice. The extra voice will make it possible for all the voices to be more elegant and expressive.

In four parts the first evil to be avoided is a stodgy overcrowded texture. The two most important features are:

(a) Lightness of texture
(b) Contrast of texture

These should be studied in the Palestrina example below. Notice that not more than five complete bars are wholly in four parts.

De – – i qui tol lis pec-ca-ta mun – –
– – i qui ___ tol-lis pec – ca – ta mun – – – – –
– gnus De – – – – – – – – i
De – – – – – i qui ___ tol-lis pec-
– – – di, qui tol – lis pec-ca – –
– – di, qui ___ tol – lis pec – ca-ta mun – – – di
qui tol – lis pec-ca – – ta mun – – – –
-ca – ta mun – – di, qui tol – lis pec-ca – – ta mun –
– ta mun – – di, mi – se – re-re no – bis, mi –
mi – se – re – re no – bis, mi – se – re-re no –
– di, ___ mi – se- re – re no – – bis, mi – se-
– di, mi – se – re-re no – bis, mi – se- re-re no – bis, ___

The exposition of four parts

The "Agnus Dei" above Ex. 161 is an example of normal practice. Note that the second entry is after one bar, the third after three bars, and the fourth after four-and-a-half bars; this lack of symmetry is intentional and a perfectly regular exposition is usually avoided.*

The "Kyrie" Ex. 162 shows three voices entering closely (note the shortening of the first note of the treble entry). The bass entry is enhanced by the delay, and the other parts have time to move into position for a suspension as it enters.

---

*I do not want to give the impression that lack of symmetry is a negative characteristic - of all periods the sixteenth century is the one in which the sense of *proportion* was most perfectly developed, and proportion is in a very real sense the opposite of symmetry - thus Ruskin: "Wherever Proportion exists at all, one member of the composition must be either larger than, or in some way supreme over, the rest. There is no proportion between equal things. They can have symmetry only, and symmetry without Proportion is not composition" (*Stones of Venice,* IV, XXVI).

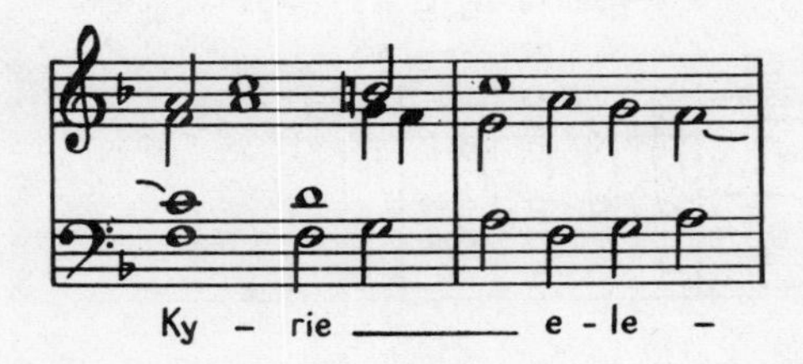

The "Kyrie" Ex. 163 shows the entry of the four voices in two pairs, producing an effect of antiphonal response:

Entry in thirds is another way of pairing off the voices. It is less used; there is a loss of independence, and in spite of the sweetness of the sound the effect is a little foursquare:

The entry of two themes is an effective device, used especially when one of the themes is a plainsong. It produces an elaborate interweaving of parts if carried out fully with each part taking both themes in succession:

A contrapuntal technique often used in setting a plainsong cantus firmus (or in the next century in writing chorale prelude) is the inversion and/or diminution of the subject. When the theme is suited to this treatment the result is effective and logical, but it can with ease produce the most deplorably dry and academic effect:

As in three parts, entry in stretto is possible but quite rare:

In the next example, Lassus shows clearly that the symmetrical entry of three voices at the same pitch produces a doleful and poverty-stricken sound:

The antiphonal grouping of voices

Undoubtedly the most important element that enters into polyphonic music at this point is the contrast of texture mentioned above. This can be seen at its furthest point of development in works for double, triple, or quadruple choirs of voices and instruments by Venetian composers such as Willaert, Cipriano da Rore, Andreas and Giovanni Gabrieli,* etc. and the principle is rooted in the ancient practice of antiphonal psalm-singing. It survives still in the Decani and Cantoris of the English cathedral choir.

A straightforward response to the statement of the two lower voices is shown in the next example. Notice that the response is not mere repetition and that the two statements overlap:

*And, of course, in the great motet "Spem in alium" by Tallis, for eight five-part choirs.

As one explores further, so an infinite variety of different vocal groupings begin to appear even from a four-voice choir. The next example shows two three-part groups:

Ex. 170 Victoria: Mass 'Quam pulchri sunt'

However, it is impossible even to begin to catalogue all the different possibilities. Voices can be used in different registers, they can cross, and so one voice may perform many different functions, as treble, bass, or middle part. In the work of the best composers we get an impression of a subtle and refined orchestration, a kaleidoscopic change of colour (albeit in pastel shades) which counteracts the monotonous texture of unaccompanied vocal music.

It is in setting the Credo and Gloria that most composers take advantage of a fully antiphonal style. In both cases the text is long and must be covered in a fairly short space of time. There is no possibility of dwelling on the words long enough to allow a completely fugal treatment, so the style is bound to be basically homophonic. An entirely homophonic setting would be unthinkably dull and the solution is found in a mixture of counterpoint and harmony, and dominating throughout, the idea of the antiphonal play of voices.

te, glo - ri - fi - ca - mus ____ te. Gra - ti - as a - gi - mus
te, glo - ri - fi - ca - mus te. ____ Gra - - ti - as a - gi - mus
te, glo - ri - fi - ca - mus te. ____ Gra - - ti - as a - gi - mus
te, glo - ri - fi - ca - - mus te. Gra - - ti - as a - gi - mus
ti - bi Do - mi - ne De - us,
ti - bi Do - mi - ne De -
ti - bi prop - ter mag - nam glo - ri - am tu - - am,
ti - bi prop - ter mag - nam glo - ri - am tu - am,
Rex coe - les - - tis, De - - us Pa - ter om - ni - po - tens. ____
- us, Rex coe - les - - tis, De - - us Pa - - ter om - ni - - po - tens.
De - - us Pa - ter om - - - ni - po - tens. ____
De - us Pa - ter om - ni - po - tens. ____

Ex. 1. Skeleton exercises of the type previously encountered in this book may be prepared from any available four-part polyphonic music.

Ex. 2. Write a short mass (i.e. Kyrie, Sanctus, Benedictus, Agnus Dei) basing as much of the thematic material as possible on a plainsong.

APPENDIX ONE

SYNOPSIS OF RULES

Any student who is prepared to take the trouble to search far enough and long enough will find it possible to quote exceptions to practically all the rules which I list below. It is an excellent, rather slow, game. It is to be hoped, however, that by the time the student is able to write in the sixteenth-century idiom, the idea of rules as oppressive restrictions will have receded in his mind to be replaced by a picture of their significance as laws of music, valid for their place and time, and rarely, if ever, absolute prohibitions.

Melodic line

(1) Intervals must be easily sung. Hence all diminished or augmented intervals are forbidden, the seventh and all intervals exceeding an octave are forbidden, and the sixth is forbidden except for the minor sixth ascending. The rule does not apply so strictly to the interval occurring between the end of one phrase and the beginning of another; this is known as a "dead interval", and the ascending major sixth and occasionally even the seventh may occur in this way.

(2) The faster the movement the fewer the leaps. A leap involves a change of vocal register and this requires time if it is to be made stylishly.

(3) After an upward leap, which causes tension, it is natural to relax by moving downward. The same applies to a lesser extent to a downward leap.

(4) Upward leaps may not be made from accented crotchets.

(5) Crotchets may not be repeated except in the anticipation. Quavers may not be repeated.

Harmonic combination

(1) With very few exceptions, and suspension dissonances apart, there must be a consonance on every minim beat.

(2) Consonant intervals in two parts are: unison, third (major and minor), perfect fifth, sixth (major and minor), and octave, and extensions of these intervals.

Consonant combinations of three notes are: third plus fifth, or third plus sixth.

(3) Adjacent parallel fifths and octaves are forbidden. Fifths may occur freely separated by one note, octaves in less than four parts should be more widely separated.

(4) Exposed fifths and octaves should be avoided in two-part work. In three parts they are permissible if the third part moves in the opposite direction.

(5) Parallel movement should be avoided as much as possible whatever the interval concerned. But see Ex. 131.

(6) Sharpened notes and leading notes may not be doubled.

(7) A piece may not end on a minor chord. It will preferably begin on a major chord, if it is in homophonic style. In two parts the piece will end on the octave or unison.

(8) Passing dissonance should only occur on unaccented beats; minim passing notes may be used in two-part work with a cantus firmus but not in two-part fugal style. If they are used in three and four parts they should move downward. They must not produce a percussed dissonance.

(9) Passing dissonances must be approached and left by step. Leaps may only be made to and from concords.

(10) Suspensions are the only discords allowed on the accented beat. They must resolve downward one step.

(11) A suspension should normally be used at cadence points.

Rhythm

(1) Rhythm should not be repetitive. Hence no sequence is allowed.

(2) Rhythm should not be symmetrical. It must be proportioned - see footnote, p. 116

(3) Rhythm should not be continuously duple - or triple, for that matter; the two metres must fluently interpenetrate one another, by means of syncopation, word placing, etc.

(4) The available note values are the breve and dotted breve, the semibreve and dotted semibreve, the minim and dotted minim, the crotchet and quaver; their equivalents in tied notes may be used.

(5) A note of small value should not normally be tied to a note of larger value, except where the second is the last note of the piece.

(6) The suspension and resolution should be of the same length, i.e. normally a minim. The preparation may be longer but not shorter.

(7) Quavers should occur in groups of two, on an unaccented beat.

APPENDIX TWO

SEMIBREVE TIME AND CROTCHET TIME

Throughout this book the unit of time has been the minim. Indeed, in one case the time values were doubled in order to bring the example into line with the others (Ex. 154). In music for the Church written in the sixteenth century, the minim beat is almost universal. But there is evidence before our eyes that this notation was not always standard. The word breve indicates a past time when it was a comparatively short note, and indeed time values have been slowing down throughout the course of musical history. A glance at a page of a modern score shows that the process is not yet halted.

In the sixteenth century there were traces of an earlier tradition which used the semibreve as the unit. It could be used for making particularly solemn or grand effects, and as a convenient resource at a cadence when one part holds a long sustained note (pedal point, in modern nomenclature), since the minim would be used as a passing note, rising or falling, and the semibreve as a suspension.

Ex. 172
Palestrina: Mass 'Papae Marcelli'
(final section of Credo)
(a)
A - - - - - - men, A - - - - - - - - - men,
A - - - - - - - - - - - - men, A - - - -
A - - - - - - - - - - men, A - - - -
A - - - men, A - - - - - - - men, A - -
A - - - - - men, ___ A - - - - - - - - - -
A - - - - - - - - men, A - - - - - men,
A - - - - - - - - - - - - - men
men
men, A - - - - - -
men
men,
A - - - - - - men, ___

In the madrigal, in instrumental music, and in any other consciously "modern" style, the unit will be the crotchet. The time signature will be 4/4 and the semiquavers will appear. However, in moments of excitement it is quite possible that the sacred 4/2 style will leap into this "crotchet time" making a "doppio movimento" effect. In such a case, crotchets will carry separate syllables and be suspended, just as in a madrigal. The time signature does not change in such a case.

3/2 time also appears, and is generally very light and uncomplicated.

* See chapter 14, page 103, triple rhythm

APPENDIX THREE

DOUBLE COUNTERPOINT

Double counterpoint is counterpoint which can be inverted; that is, the bass and treble can change places. It would be more clearly described as 'invertible counterpoint', and it is unfortunate that the term 'inversion' has to serve for another device. Where there is any doubt, indicate 'inversion of position' or 'inversion of direction'.

Double Counterpoint in the Octave

This is the simple move of an octave up or down, as in the following example. The student will have noticed a tendency in two-part exercises with a cantus firmus to write the same counterpoint above and below. Double counterpoint in the octave legalizes this rather lazy technique.

Even in this straightforward process it is clear that the intervals are changed by the inversion. The table gives the relationship.

intervals above - 1 2 3 4 5 6 7 8 become

intervals below - 8 7 6 5 4 3 2 1 ------ and vice versa

From this it can be seen that;

(a) The octave, unison, third, and sixth, remain consonant in inversion.

(b) The seventh inverts to become a second, and the second to a seventh: so that if they are correctly treated as suspensions or passing notes they remain correct in inversion.

(c) The fifth inverts to become a fourth.

The following rule then, governs double counterpoint in the octave:

The interval of the fifth must be avoided, or treated as if it were a dissonance (see above Ex. 175*)

Some practice should be had in writing double counterpoint in the octave, above and below a plainsong, and also with both parts in the free "fifth species" style.

It should be observed, however, that invertible counterpoint in the octave plays little or no part in sixteenth-century compositions. At the most, tiny fragments of it can be found in the Gloria or Credo of a mass, as in the following examples, although Ex 176 b is in fact not true double counterpoint since one note has to be covered by a sustained bass. In an age when variety of texture and a constant development of the musical idea was the ideal, this calculated contrapuntal thinking would be quite out of place, although the technique was in fact well known. It was not until the next century that the idea of a counter-subject in double counterpoint with the subject was to play an important part in the structure of the Baroque Fugue.

Double Counterpoint in the Twelfth

This is incomparably the most fruitful of the various types, and fairly frequent examples of its use in the fugal exposition of two themes can be found:

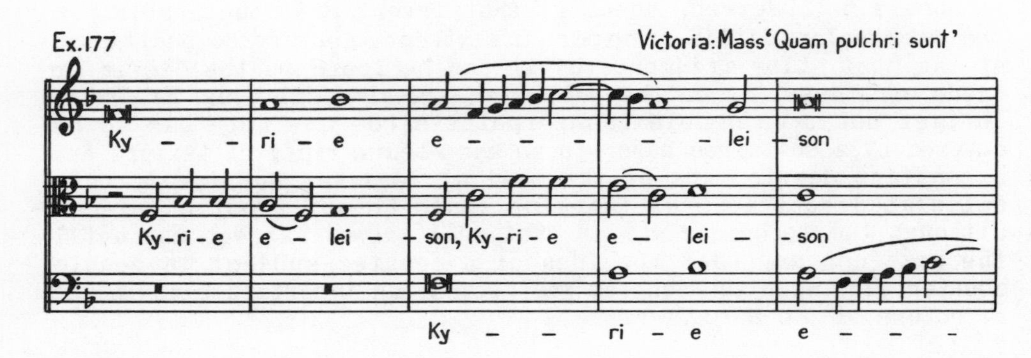

It has a structural and tonal significance not possessed by inversion in the octave, nor indeed by any of the other inversions.

It is probably intentional that in the duo which Orlando di Lasso writes to demonstrate the multiplicity of devices than can be used to manipulate a simple Cantus Firmus, he begins with an inversion in the twelfth, and omits the octave.

The alteration of intervals is shown by the following table:

1	2	3	4	5	6	7	8	9	10	11	12
12	11	10	9	8	7	6	5	4	3	2	1

The following rules arise from this:

(1) The interval of the sixth must be treated as if dissonant, i.e. either completely avoided, or treated as a crotchet passing note.

(2) The 7:6 suspension is unavailable.

Double counterpoint in the tenth

In this case the table:

1	2	3	4	5	6	7	8	9	10
10	9	8	7	6	5	4	3	2	1

shows that all consonances remain consonant, but that thirds become octaves, sixths become fifths, and vice versa.

It follows that all similar motion is impossible since it will produce parallel fifths or octaves.

The rule in this case is:

Parts must proceed in contrary motion.

If this rule is kept then it is possible to double the counterpoints as shown in tenths (or of course in thirds). This is perhaps the most popular function of this inversion.

Finally, this fragment from Byrd's Motet "Of flattering speech" shows the economy of material that is possible by using subjects in double counterpoint.

APPENDIX FOUR

(g)
Urbs be-a-ta Je-ru-sa-lem Dic--ta pa-cis vi-si-o
Quae con-stru-i-tur in coe-lis Ni-vi ex la-pi-di-bus
Et an-ge-lis co-ro-na-ta Ut spon-sa-ta co-mi-te.
(h)
Pan-ge lin-gua glo-ri-o--si Cor-por-is my-ste-ri-um
San-gui-nis-que pre-ti-o-si Quem in mun-di pre-ti-um
Fruc-tus ven-tris ge-ner-o--si Rex ef-fu-dit gen-ti-um.
(i)
I--ste Con-fes-sor Do-mi-ne sa-cra-tus Fes-ta plebs cu-jus ce-le-brat per or-bem
Ho-di-e lae-tus me-ru-it se-cre-ta Scan---de-re coe-li.
(j)
Sanc-to-rum me-ri-tis in-cly-ta gau-di-a Pan-ga-mus, so-ci-i, ges-ta-que for-ti-a:
Num glis-cit a-ni-mus pro-me-re can-ti-bus Vic-to-rum ge-nus op-ti-mum.

INDEX

Accented note 11

Anticipation 44

Antiphonal response 119ff.

ANSWER 20, 57ff.

Answer by diminution and inversion 118

Tonal answer 58

Augmented intervals 5

Augmented fourth (and see 'Tritone') 76

Auxiliary note 24

Bass in three-part writing 78

CADENCE 19, 42

Final cadence 54

Harmony at cadence 46

Modal cadence 53

Trills at cadence 108

Canon 17, 70

Chords 75ff.

Consecutives 5, 12ff, 26, 31, 76, 124.

Consonances, perfect 13

Consonance and dissonance 5, 29ff. 46, 56ff, 75, 124.

Consonant fourth 87

Contrary motion 13, 31, 32, 77

Countersubject 90, 102

CROTCHET

Crotchets in fifth species 41
Crotchet idioms 27
Crotchets, third species 24
Crotchet time 125ff.

Crossing(in three parts) 77

Dead interval 123

DECORATION

Anticipation 44
Decorated suspension 44ff.
Quaver decoration 46
Third species 24ff.

Diatonic seventh chord 85
Diminished fifth 76, 85
Dissonance 5, 29ff, 46, 56ff, 75, 124
Dissonance, percussed 29
Dotted rhythm 17
Double counterpoint 129(App. III)
Doubling 77

Exposed fifths and octaves 6, 27, 31, 77,

EXPOSITION

Countersubject exposition 90, 102,
Four-part exposition 116ff.
Imitative entries 57ff.
Stretto exposition 90, 102
Tonal answer 58

Faux bourdon 76

FIFTHS

Exposed fifths 6, 27, 31, 77,
Parallel fifths 5, 12ff, 26, 31, 76, 124

Harmonic resources 76

HARMONY 5ff, 46

Homophonic style 73ff.
Modal cadence 53

IDIOMS 61ff.

Consonant fourth 87
Nota cambiata 61
Third species idioms 27

Imitation 9, 20, 57ff.
Instrumental music 104ff.
Invertible counterpoint - see Double counterpoint.

MELODIC LINE 37ff, 98, 123

Plainsong 2
First species 5
Third species 25

MINIM

Passing note 12
Restriction of passing note 15, 57
Tied minims 17

Modes 1, 50ff.
Motet 56, 98
Musica ficta 50

Nota cambiata 61

Octaves

Exposed octaves 6, 27, 31, 77
Parallel octaves 5, 12ff, 26, 31, 76, 124

Passing note 15, 17, 24
Percussed dissonance 29
Perfect consonances 13, 15
Plainsong 1ff.
Plainsong cantus firmi 135
Plainsong fantasia 82

PROPORTION
  Proportion and symmetry 116
  Rhythmical proportion 80

Quavers 45, 57

Rests 20
Repeated minims and semibreves 39
Resolution 29

RHYTHM 3, 6, 17ff, 38ff.
  Dotted rhythms 17
  Quavers 45
  Syncopation 17ff, 28ff.
  Triple rhythm 80ff, 103ff, 125

Semibreve time 125ff.
Shadowing 13
Spacing (three parts) 77
Stretto 90, 102

SUSPENSIONS 19, 28ff, 42, 44, 60, 78, 85, 87, 88ff.
  Decorated suspensions 44ff.
  Double suspensions 88
  Resolution 29
  Suspension in triple time 103

Syncopation 17ff, 28ff, 39
Symmetry and proportion 116

Tactus 36
Tempo 7, 80, 104
Texture 73ff, 114, 119
Ties 40
Time 125
Tonal answer 58
Transposed modes 54
Tritone 3, 25
Triple rhythm 80ff, 103ff, 125.